Agrimonde – Scenarios and Challenges for Feeding the World in 2050

Sandrine Paillard • Sébastien Treyer • Bruno Dorin
Editors

Agrimonde – Scenarios and Challenges for Feeding the World in 2050

Springer

Editors
Sandrine Paillard
ANR
Paris
France

Bruno Dorin
CIRAD
Montpellier
France

Sébastien Treyer
IDDRI
Paris
France

Éditions Quæ, RD 10, 78026 Versailles Cedex, France

ISBN 978-94-017-8744-4 ISBN 978-94-017-8745-1 (eBook)
DOI 10.1007/978-94-017-8745-1
Springer Dordrecht Heidelberg New York London

Library of Congress Control Number: 2014932834

Printed on acid-free paper

Springer is part of Springer Science+Business Media (www.springer.com)

Preface

Could we feed the whole world today? Strictly speaking yes, from the point of view of the quantities of food produced by agriculture (3,000 kcal/cap/day in 2003)—but this of course fails to take into account crises and wars, inequalities, speculation and unaffordable prices for the rural and urban poor, loss and waste, climate-related accidents and pest invasions which starve peasants in many parts of the world. Amartya Sen, in his book "Poverty and Famines" published in 1981, puts it in a nutshell: "Starvation is the characteristic of some people not *having* enough food to eat. It is not the characteristic of there *being* not enough food to eat"; hence the billion undernourished people in the world in 2009. What about in 2050 when 3 billion more human beings are expected to be living on this planet? This is the crucial question considered in this book; a question on which INRA and CIRAD decided to work together by launching the Agrimonde foresight project in 2006.

Although it may sound straightforward, in practice the question is difficult to address. The food and agricultural challenge cannot be limited to theoretically satisfying the nutritional needs of the earth's population in quantitative terms only. It is also a matter of enabling everyone to have access to sufficient food that is safe from a sanitary point of view as well as being nutritionally balanced. Nor can the challenge be reduced to a basic equation of supply and demand of agricultural goods and food. Agricultural and food systems also have to be grounded in a logic of sustainable development, and to take into account the problem of energy induced by the gradual depletion of fossil fuel reserves, not to mention the growth of social inequalities. As well as producing more, the world's farmers will also have to produce more efficiently, using practices and systems that make sparing use of fossil fuels and natural resources. Additionally, they will have to produce other things: energy and industrial goods as substitutes for petrochemical products, as well as environmental and territorial services (soil and water preservation, biodiversity protection, carbon storage, prevention and limitation of fires and floods, provision of open and diversified landscapes).

In view of the complexity of the question, we decided to adopt a foresight approach in which we first considered two scenarios describing possible, sharply contrasting futures. Both have the same timeline (2050) and the same assumptions of demographic growth in each large region of the world and of migrations between regions. They differ however in their trajectories in the evolution of food and agricultural systems—trajectories that represent two contrasting visions of tomorrow's world. The first scenario represents a continuation of current trends in the production and use of various types of food biomass in a "liberalised" world where the priority is economic growth and the material well-being of current generations. The second trajectory sets the objective of satisfying global food needs, amounting to 3,000 kcal per capita per day, including 500 of animal and aquatic origin, in all the regions of the world. This implies less consumption and waste in developed countries and substantial increases in food consumption in many developing countries. Here the objective underlying the evolution of the food and agricultural systems of the world's regions is sustainability.

This book describes the construction and analysis process of these two scenarios, along with the conclusions that can be drawn from them. Without going into any detail here, three challenges are highlighted by this foresight exercise, concerning: diets and their impacts on major balances; technological and organisational choices in agricultural production and the feasibility of an agriculture that is both intensive and ecological; and international trade in agricultural and agri-food products, including the possibility of making it secure on a global scale. These three challenges are considered in the last chapters of the book.

This initial endeavour calls for follow-up in two respects: first, the setting up of a permanent, quantitative and qualitative platform in France that could serve to reinforce reflection on the future of food and agriculture; and, second, the identification of priority research questions put to international agronomic research. A platform for this purpose will be created by CIRAD and INRA at the end of 2010.

Before leaving you to enjoy this book, we wish to thank everyone who participated in this initial phase of the "Agrimonde adventure": the scientists from INRA, CIRAD and other research and education institutions, as well as all the experts who were asked to contribute and who offered us not only their advice and criticism but also their encouragement.

President of INRA — Marion Guillou
President of CIRAD — Gérard Matheron

Agrimonde: A Work Collective

Under the responsibility of a CIRAD-INRA steering committee, the Agrimonde scenarios were developed, analysed and discussed by a collective consisting of a panel and a project team which met through 20 workshops from June 2006 to December 2008.

Steering Committee
Patrick Caron, CIRAD (French Agricultural Research Centre for International Development)
Catherine Esnouf, INRA (French National Institute for Agricultural Research)
Hervé Guyomard, INRA
Bernard Hubert, FI4IAR (French Initiative for International Agricultural Research)
Alain Weil, CIRAD
Project Team
Sébastien Treyer (Project Leader), IDDRI (Institute for Sustainable Development and International Relations)
Maryse Aoudaï, INRA
Jean-Marc Chaumet, INRA
Bruno Dorin, CIRAD
Tristan Le Cotty, CIRAD
Sandrine Paillard, INRA
Tévécia Ronzon, INRA
With the collaboration of Rémi Barré, INRA and CNAM, Isabelle Karcher, INRA and Laurent Parrot, CIRAD
Expert Panel
Bernard Bachelier, FARM (Foundation for World Agriculture and Rural Life)
Danielle Barret, CIRAD
Pierre-Marie Bosc, CIRAD
Jacques Brossier, FI4IAR
Jean-Pierre Butault, INRA
Jean-Christophe Debar, AGRI US Analyse
Francis Delpeuch, IRD (Research Institute for Development)

Fabrice Dreyfus, SupAgro Montpellier (International Center for Higher Education in Agricultural Sciences, Montpellier)
Gérard Ghersi, MSH Montpellier (The Montpellier Social Sciences and Humanities Research Institute)
Michel Griffon, CIRAD
Christian Hoste, CIRAD
Bernard Hubert, FI4IAR
Denis Lacroix, IFREMER (French Research Institute for Exploitation of the Sea)
Marie de Lattre-Gasquet, CIRAD, ANR (French National Research Agency)
Jacques Loyat, MAP (French Ministry of Agriculture and Fisheries)
Michel Petit, IAM (Mediterranean Agronomic Institute of Montpellier)
Jean-Louis Rastouin, SupAgro Montpellier (International Center for Higher Education in Agricultural Sciences, Montpellier)

Introduction

Bruno Dorin, Sandrine Paillard and Sébastien Treyer

The interactions between agricultural production, food and sustainable development are the subject of a growing number of debates and studies, in which different visions of the world and different positions in the geopolitical arena are revealed and brought into confrontation. These interactions concern issues of wealth distribution and international regulations on global trade and public goods. They also relate to conceptions of the science-society nexus, of progress, of our planet's future and humans' role therein.

These debates and analyses have been enhanced over the past few years by the confluence of research of diverse origins and traditions: econometric modelling dedicated to the analysis of agricultural or energy policies; global foresight studies of geopolitical inspiration or concerned by global sustainable development; modelling of the impacts of climate change; and, more recently, studies stemming from international scientific assessment on ecosystems or biological diversity. There are very few countries, international organisations, major NGOs, or leading firms today that do not, in one way or another, contribute to the debate on the world's future agricultural production and food, their interactions with the objective of sustainable development, and its implications for international relations, public policy and research priorities.

On the whole, this abundance of information, data and approaches fails to provide a coherent picture. The new awareness of global risks has led to a proliferation of analyses and of the international arenas in which they are presented, but has not necessarily clarified the challenges and options available to address them. Nor has it prompted many questions on the underlying assumptions structuring the debates, or brought to the fore approaches out of line with international *doxa*.

By identifying leeway through new reflection opportunities, foresight studies seek to further the evolution of these debates. The challenge is immense since these studies and the debates they generate are the crucible where many concepts and scientific and political arguments are converted into standards. These, in turn, will weigh on international negotiations on agriculture, international trade and development aid, as well as on the action of multilateral organisations.

The CIRAD-INRA Agrimonde initiative stemmed from the desire to promote reflexivity and interaction. Stimulated by the work of the Millennium Ecosystem

Assessment (MA) (MA, 2005a) and the International Assessment of Agricultural Knowledge, Science and Technology for Development (IAASTD) (IAASTD, 2009), this initiative defined three main objectives: (1) to design the modalities of strategic reflection based on a scenario approach, in order to guide research orientations in the field of agronomy and food, broadly speaking; (2) to initiate a process of debates, interactions and appropriation of these topics on a national scale; (3) to favour the participation of French experts in international debates on the subject.

The Agrimonde foresight study has acted as a platform for the analysis and discussion of scenarios. It was designed as a mechanism of interaction between "experts": researchers, decision-makers and, more generally, stakeholders and actors within the system under consideration.

The platform was also designed to function according to the basic principles of a foresight approach, i.e.:

- consideration of the systemic nature of reality by integrating the multiplicity of relevant variables as far as possible,
- explicit expression of divergences, including the variety of worldviews and of scientific and institutional positions,
- integration of scientific uncertainties by revealing the assumptions underlying alternative scenarios, with a view to exploring possibilities and not to making predictions,
- collective learning that impacts on each actor's representations by involving experts and stakeholders in the work itself,
- reflection with a long-term view in order to assess individual and combined effects of the evolution of variables, as well as the impacts of assumptions on possible futures,
- transparency of the work progress, reference to the best scientific studies to constitute analytical and databases, and clarification of the simplifications and assumptions made.

This book presents the results of work undertaken by the Agrimonde Expert Panel during the period from 2006 to 2008. The first two chapters present the foresight platform designed for scenario-building and, in particular, the quantitative Agribiom module which was the main tool used to quantify the scenarios. Chapter 3 is devoted to a review of the food economy over more than four decades (1961–2003), on a global scale and in six major regions of the world (Middle East—North Africa, Sub-Saharan Africa, Latin America, Asia, the former Soviet Union, and OECD-1990). Knowledge of past trends has proved indispensable in making assumptions on the evolution of agricultural resources and uses in the period up to 2050. Chapter 4 explains the choice of the two scenarios: one is essentially a trend-based scenario (Agrimonde GO) while the other shows a distinct break with previous trends and foresees sustainable food and agriculture in 2050 (Agrimonde 1). Chapters 5, 6, 7, 8, 9 and 10 present the successive stages of the scenario-building process. First the quantitative assumptions on food uses are described (Chap. 5), followed by those on agricultural resources (agricultural areas in Chapter 6 and food crop yields in Chap. 7). The food resource-use balances are then calculated (Chap. 8) to assess the conditions under which the world and its main regions feed their populations in each of the scenarios. The coherence of the scenarios, their comparison and the driving forces underlying them are analysed in Chap. 9. Chapter 10 proposes an account of complete scenarios, combining quantitative and qualitative assumptions. Finally, Chap. 11 invites the readers into the debate by putting forward three points of view on the results of Agrimonde, with regard to the evolution of diets, ecological intensification and trade.

Acknowledgments

The Agrimonde collective is very grateful to Kais Abbes, Gilles Allaire, Rigas Arvanitis, Olivier Clément, Paul Colonna, Yves Dronne, Christophe Ferlin, Louise Fresco, Vincent Gitz, Ghislain Gosse, François Houllier, Jean-Charles Hourcade, Géraldine Kutas, Xavier Leverve, Claudia Ringler, Laurence Roudart, Christian Salle and Frank Schmitt for their contribution to its work. It would also like to thank the French Energy Council (*Conseil français de l'énergie*) for its support, and Jean-François Foucher and Claude Ladhuie for their technical and logistic assistance.

Contents

Contributors

Patrick Caron CIRAD, Montpellier, France

Jean-Marc Chaumet Institut de l'Elevage, Paris, France

Tristan Le Cotty CIRAD—UMR CIRED, Montpellier, France

Francis Delpeuch IRD, Rabat Agdal, Maroc

Bruno Dorin CIRAD—UMR CIRED, Montpellier, France

Gérard Ghersi Montpellier University, Montpellier, France

Hervé Guyomard INRA, Paris, France

Bernard Hubert Agropolis international, Montpellier, France

Sandrine Paillard INRA, Paris, France

Michel Petit IAMM, Montpellier, France

Jean-Louis Rastoin SupAgro Montpellier, UMR 1110 Moisa, Montpellier, France

Tévécia Ronzon INRA, Paris, France

Sébastien Treyer IDDRI, Paris, France

Abbreviations and Acronyms

Names of the quoted scenarios

AGO	Agrimonde GO
AG1	Agrimonde 1
AM	Adapting Mosaic
BAU	Business-As-Usual
CRI	Water Crisis
DGR	Doubly Green Revolution
GO	Global Orchestration
OS	Order from Strength
SUS	Sustainable Water Use
TG	TechnoGarden

Designation of the world's regions

ASIA	Asia
FSU	Former Soviet Union
LAM	Latin America
MENA	Middle East - North Africa
OECD	OECD-1990
SSA	Sub-Saharan Africa

Designation and selected variables in the Agrimonde foresight

AEZ	Agro-Ecological Zones
GAEZ	Global Agro-Ecological Zones
LUT	Land Utilisation Types
SUA	Supply Utilization Accounts
AQUA	Fresh-water products
MONO	Non-grazing or monogastric animals (Agribiom biomass category)
MARI	Marine products (Agribiom biomass category)
RUMI	Grazing animals or ruminants and large herbivorous animals (Agribiom biomass category)
FCA	Food cultivated area
NFCA	Non-food cultivated area
VEGE	Plant (Agribiom biomass category)

AREA	Area (ha)
TRAD	Net trade balance (total exports—total imports) of foodstuffs (kcal)
ΔSTOC	Stock variation of foodstuffs (kcal)
POPU	Human population (inhabitants)
FOOD	Quantity of foodstuffs (kcal) used for feeding the human population, including wastage occurring in the household
Feed	Quantity of foodstuffs (kcal) used for feeding animals
SEED	Quantity of foodstuffs (kcal) used for reproductive purposes (seed, eggs for hatching, etc.)
VANA	Quantity of foodstuffs (kcal) used for non-food purposes: lubricants, energy, cosmetics, biomaterial, etc.
WAST	Wasted quantity of foodstuffs (kcal) between the general available quantities (Production – Exports + Imports +/– Stocks) and their allocation to a specific use (food, feed, etc.); this does not include losses occurring before and during harvesting, and wastage occurring in the household

Unitscap	Capita
d	Day
g	Gram
G	Giga, billion
Gkcal	Giga kilocalorie
ha	Hectare
k	Kilo, thousand
kcal	Kilocalorie
kg	Kilogram
M	Mega, million

Acronyms

CFE	*Conseil français de l'énergie*, French Energy Council, France
CGIAR	Consultative group on International Agricultural Research
CIHEAM	*Centre international des hautes études agronomiques méditerranéennes*, International Center for Advanced Mediterranean Agronomic Studies, France
CIMMYT	International Maize and Wheat Improvement Center, Mexico
CIRAD	*Centre de coopération internationale en recherche agronomique pour le développement*, Agricultural Research Centre for International Development, France
CIRED	*Centre international de recherche sur l'environnement et le développement*, International Research Centre on the Environment and Development, France
CNAM	*Conservatoire national des arts et métiers*, France
FAO	Food and Agriculture Organization
FARM	*Fondation pour l'agriculture et la ruralité dans le monde*, Foundation for World Agriculture and Rural Life, France
FI4IAR	French Initiative for International Agricultural Research, France

GMBD	Global Maritime Boundaries Database
GPAFS	Global Partnership for Agriculture and Food Security
IAASTD	International Assessment of Agricultural Knowledge, Science and Technology for Development
IDDRI	*Institut du développement durable et des relations internationales*, Institute for Sustainable Development and International Relations, France
IFPRI	International Food Policy Research Institute
IFREMER	*Institut français de recherche pour l'exploitation de la mer*, French Research Institute for Exploitation of the Sea, France
IIASA	International Institute for Applied Systems Analysis
IMWI	International Water Management Institute
INRA	*Institut national de la recherche agronomique*, French National Institute for Agricultural Research, France
IPCC	Intergovernmental Panel on Climate Change
IRD	*Institut de recherche pour le développement*, Research Institute for Development, France
IUCN	International Union for Conservation of Nature
MA	Millennium Ecosystem Assessment
NGO	Non-Governmental Organization
OECD	Organisation for Economic Co-operation and Development
SCAR	Standing Committee on Agricultural Research
UN	United Nations
UNDP	United Nations Development Programme
USDA	United States Departement of Agriculture
WHO	World Health Organization
WTO	World Trade Organization

Chapter 1
Agrimonde: A Platform for Facilitating Collective Scenario-Building

Sébastien Treyer, Sandrine Paillard and Bruno Dorin

Between 2006 and 2008, the Agrimonde Expert Panel focused on building two scenarios. In the first, "Agrimonde 1", we imagined a food and agricultural system designed to be sustainable by 2050. The aim was to explore the conditions required for achievement, the dilemmas and the main challenges that such a scenario entails. "Agrimonde GO", the second scenario, is more trend-based. It consists of an application of the Millennium Ecosystem Assessment (MA) Global Orchestration scenario (MA 2005b) into a food and agriculture scenario, whereas the MA built this scenario as a description of one of the possible futures of ecosystem management.

The development of these two scenarios served as a "prototype" for the conception of the scenario-building tool presented in this chapter. In the first part of the chapter, the scenario-building methodology is described. The second part analyses its position compared to other approaches to long-term food balances.

A Tool Based on the Complementarity of Quantitative and Qualitative Analyses

Agrimonde is based on the collective work of an Expert Panel formed to bring together various points of view and opinions conveying a variety of visions of the future, so as to collectively build contrasting scenarios.

S. Treyer (✉)
IDDRI, 27 rue Saint Guillaume, 75337 Paris Cedex 7, France
e-mail: sebastien.treyer@iddri.org

S. Paillard
ANR, 212 rue de Bercy, 75012 Paris, France
e-mail: sandrine.paillard@agencerecherche.fr

B. Dorin
CIRAD—Umr CIRED, TA C56/15, 73 rue J.F. Breton,
34398 Montpellier Cedex 5, France
e-mail: bruno.dorin@cirad.fr

S. Paillard et al. (eds.), *Agrimonde – Scenarios and Challenges for Feeding the World in 2050*,
DOI 10.1007/978-94-017-8745-1_1, © Éditions Quæ, 2014

The Agrimonde platform was designed to facilitate collective analysis of the challenges facing the world's food and agricultural systems. These can be summed up as the key challenge of adequately feeding a population of nine billion individuals in 2050, while preserving the ecosystems from which other services are also expected: bio-energies, biodiversity, carbon storage, climate regulation, etc. The variables to consider when analysing this question are multifarious: geopolitical, social, cultural, sanitary, economic, agronomic, ecological, technological, etc. Moreover, the global scale on which this question is raised does not preclude reflection at regional level, for the diversity of the world's food and agriculture, and their interactions through trade and global environmental change, are key variables for the future.

Considering the number and diversity of variables, as well as the importance of the articulation between regional and global scales, classical scenario-building methods had to be adapted[1] by building a platform essentially based on the complementarity of quantitative and qualitative analyses. The formulation of sets of quantitative assumptions, at regional level on a limited number of variables (those of the Agribiom quantitative module: Chap. 2), enabled the panel to reduce the complexity while providing an entry point for in-depth qualitative analysis covering all the dimensions of the system. These dimensions were structured on the basis of a morphological analysis of the food and agricultural system. The resulting framework (Table 1.1) highlights the systemic nature of global food and agriculture; it is an important tool for testing and constructing the coherence of scenarios.

The Agrimonde scenarios were built in three main stages (Fig. 1.1):

- choosing the scenarios and the principles underlying their construction,
- building the quantitative scenarios,
- completing the quantitative scenarios with qualitative assumptions on the evolution of other dimensions and variables.

The panel first chose scenarios to build and their underlying principles, that is, their salient features (e.g. "Do we wish to build a scenario based on past food consumption trends, or a rupture scenario?"; "Do we wish to build a scenario characterised by an energy crisis?"). The timeline and geographic zoning into regions were defined at this stage, in relation to the objectives of the foresight study. In order to facilitate dialogue between the Agrimonde project and the MA scenarios, the timeline and geographical zoning of the latter were chosen for the Agrimonde 1 and Agrimonde GO scenarios.

The panel then formulated quantitative assumptions by giving a value to each of the variables within the timeline chosen and for each of the regional zones. These variables served to calculate the agricultural resources and their uses for each zone

[1] For instance, the classic French scenario method is based on a first phase of exhaustive recording of all kinds of variables liable to impact on the future of the system studied, within the timeline chosen for the foresight study. This method proved to be inappropriate since it is hardly feasible to combine assumptions on all the key variables for the future of the system studied, on both a regional and a global scale. For further details on the French scenario method see, for example, De Jouvenel (2000).

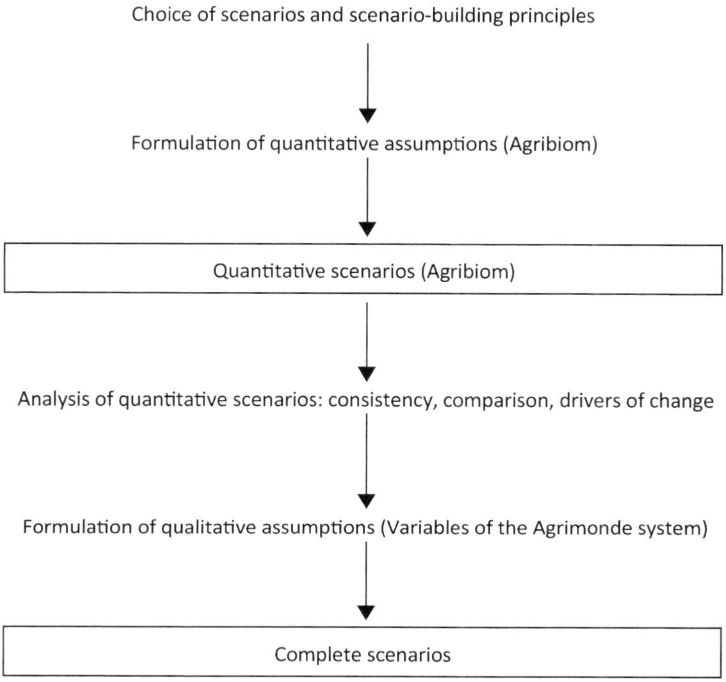

Fig. 1.1 The scenario-building phases

and for the world. For this purpose, the panel drew on its members' expertise and on retrospective analyses of past trends, as the variables to quantify were provided over several decades in Agribiom. Moreover, the panel analysed and discussed academic and foresight studies, which enabled it to systematically explore possibilities of discontinuity and, *in fine*, of contrasting futures.

To assess biomass uses, assumptions had to be made above all on the sizes of human populations, their diets in terms of calories, and the composition of those diets in terms of calorie sources (plant, land animal, aquatic plants and animals). To assess biomass resources, assumptions had to be made primarily on land use (food-crop and non-food-crop areas, pastures, forests, etc.), the productivity of cultivated areas in terms of plant calories, and the process of conversion of plant calories into animal calories.

The Agribiom module enabled us to test whether the regional resource-use ratios of food biomass (surpluses or shortages, depending on the case) balanced out at global level, including the quantities of plant biomass necessary to produce food of animal origin. When there was a global balance, assumptions were made on trade between regions. Otherwise, the panel had to revise some of its assumptions until a global balance of food biomass resources and uses (plant, animal, etc.) was obtained.

The sets formed by the quantitative assumptions and the corresponding resource-use balances constituted quantitative scenarios. The panel then completed them with qualitative assumptions. The complementarity of the quantitative and qualitative analyses operated throughout the formulation of the quantitative assumptions. When the experts formulated assumptions on diet, land use, yields or inter-regional trade, they had to analyse all their implications and ramifications. Through this process, they enhanced the basic quantitative assumptions with a set of qualitative assumptions (Table 1.1). More specifically, the quantitative scenarios were analysed for each region at global level, according to three types of question:

- Is the quantitative scenario for a particular region consistent with the scenario-building principles defined at the outset for this scenario? If not, which qualitative assumptions would make it consistent?
- What does comparison of the various scenarios teach us? Are they very different? What qualitative assumptions would be needed to ensure that they represented clearly different pictures of the future?
- What are the main challenges of this scenario? What are the main drivers of change that should be activated for it to become reality? What qualitative assumptions should be made to integrate these drivers into the scenario?

This analysis enabled the panel to consider qualitative assumptions for each of the variables of the Agrimonde system and to produce complete scenarios.

Agrimonde, Complementary to Other Approaches on Long-Term Food Balances

With the aim of exploring the long-term future of global food and agricultural systems, the Agrimonde platform was designed to be a complementary method to existing approaches, especially the following:

- Forecasting exercises centred on possibilities of increases in the global agricultural production to meet the needs of growing populations. These have regularly been carried out since the creation of the Food and Agriculture Organization of the United Nations (FAO). In such exercises, as in Agrimonde, resource-use balances on regional and global scales are discussed,
- Sectoral economic modelling exercises, like those of the International Food Policy Research Institute (IFPRI), are coupled with or integrated into other models to define, explore or further develop global scenarios such as those of the Intergovernmental Panel on Climate Change (IPCC) (IPCC 2000). These global scenarios now constitute a family of reference scenarios (World Water Vision (Cosgrove and Rijsberman 2000), Millennium Ecosystem Assessment (MA 2005b), Global Environmental Outlook (UNEP 2007), etc.). Agrimonde can contribute to discussions on these scenarios, especially by proposing as yet unexplored options concerning food and agriculture.

Table 1.1 Variables of the Agrimonde system

1. Global context		Population Urbanisation and rural exodus Economic growth Advances in knowledge Income distribution Agricultural commodity prices
2. International regulations		International political relations Organisation of international trade International agreements on climate International agreements on biodiversity Governance and management of sanitary risks Governance and management of marine resources North–South capital flows
3. Dynamics of agricultural production		Production areas Investments in farming Investments in infrastructures and public goods Social forms of production Production techniques Processing (agro-industry): organisation and production technologies
4. Dynamics of biomass consumption		Consumption habits and diets Society's awareness of sanitary issues Society's awareness of environmental issues Consumption of biomass for energy production Consumption of biomass for the production of industrial goods
5. Actors' strategies	States' strategies	Agricultural policies Sanitary and nutrition policies Energy policies Environmental policies
	Private actors' strategies	Role of professional agricultural organisations Strategies of multinational firms Role of NGOs
6. Knowledge and technologies in the field of food and agriculture		Investments in public and private R&D Objectives of innovations Intellectual property system for living organisms Orientations of agricultural research Farmers' training Organisation and actors of innovation and its diffusion
7. Sustainable development	Natural resources	Biodiversity conservation Greenhouse gas emissions and climate Soil fertility Water (availability and quality)
	Social equity	Satisfaction of essential needs (food, health, employment, education) Quality of life: dwellings, culture, social relations

A Set of Quantified Scenarios on the Future of Agriculture and on Long-Term Global Food Balances

The comparison between the biomass production potential and additional needs related to future population growth is a long-standing issue but one that is fundamental for analysing the future of agriculture. The initial Malthusian question compares the exponential growth of food needs with linear growth of agricultural yields. It has been rendered far more complex by analyses on agricultural production capacities and the role of technological progress.

The FAO and IFPRI have reviewed these analyses and highlighted two types of approaches (Mc Calla and Revoredo 2001):

- The first type of approach is qualified as "neo-Malthusian". From Meadows' report for the Club of Rome (Meadows et al. 1972) to the more recent work of Lester Brown (Brown 1995), it aims to emphasise the problems generated by a growing demand faced with the limits of available natural resources in the medium and long term,
- Other approaches aim rather to highlight the constraints to overcome in order to maintain the increase in agricultural production. Regularly updated trend projections such as those proposed by the FAO (FAO 2002) are part of this type of approach, as are assessments of the maximum potential of agricultural production (exercises such as "Wageningen Limits of Food Production" (Luyten 1995)) that we have to strive to attain by removing technical or socio-economic obstacles (Koning et al. 2008).

Within the framework of this analysis, two exercises have been decisive in the design of our Agrimonde scenarios: that of Philippe Collomb (Collomb 1999) and that of Michel Griffon (Griffon 2006). In both cases the accent is not only on covering global needs with adequate resources, but also on regional capacities to cover regional needs. In both cases, the assessment of potential to increase production aims at estimating not only regional and global capacities to meet food needs, but also agricultural systems' capacities to participate in the effort to alleviate poverty.

The Agrimonde platform is situated in the tradition of these exercises which process aggregated data for large regions of the world, and whose contribution consists in drawing conclusions from the analysis of global and regional resource-use balances. The Agrimonde platform was designed to facilitate the exploration of new scenarios which can be qualified to varying degrees as trend-based or else as rupture scenarios. This exploration is made transparent owing to the Agribiom quantitative module whose architecture and specific features are presented in Chap. 2. These quantitative assessments do not of course afford access to the territorial complexity—of a technical, organisational or socio-economic nature—of agricultural production systems in the various regions of the world. As they are not designed to represent all issues pertaining to food and agriculture, they leave open important qualitative dimensions (farming systems, territories, public policies, etc.). These would need to be studied if the possible futures of food and agriculture were to be adequately explored.

A Genealogy of Global Scenarios and Integrated Models on Natural Resources

From the four groups of IPCC scenarios, via the World Water Vision and Global Environment Outlook scenarios, to the MA, a set of global reference scenarios has been emerging. These scenarios have been built in relation to one another and with the same integrated modelling tools (IMAGE[2], IMPACT[3], WaterGAP[4]). Integrated assessment has the main advantage of articulating socio-economic scenarios of greenhouse gas emissions and their impacts on climate change, with modelling of the consequences of these impacts on the resources and activities envisaged, for example agriculture. IMPACT is a market equilibrium model for a set of key agricultural products. It is used for long-range simulations (2025, 2050) of agricultural market equilibriums and prices, and to take into account possible substitutions, based on a number of hypotheses on production capacities, mostly from FAO projections (Rosegrant et al. 2001). The IMAGE model additionally simulates changes in land use.

With these reference scenarios the main aim is to explore four contrasting images of what the socio-economic and geopolitical states of the world might be in 2050 or even 2100 (Box 4.1, Fig. 4.1). These contrasting states can be defined along two axes: the first represents worlds in which the economy is globalised, and worlds in which it is regionalised; the second represents societies' management of the natural environment, with on the one hand proactive management and, on the other, reactive management (i.e. intervention after the emergence of an environmental crisis).

These four main global scenarios are then converted into input parameters of integrated models which, among other things, ensure the coherence of the main macro-economic variables and serve to simulate the impacts of these socio-economic scenarios on the climate, ecosystems or natural resources. This approach presented under the heading "story and simulation" (EEA 2001) consists in combining the qualitative scenarios with the integrated models, to describe possible worlds.

It would be useful to implement these integrated modelling approaches so as to represent scenarios other than those belonging to the set of reference scenarios. This would make it possible to explore new questions or issues specific to a particular

---

[2] IMAGE is an integrated model for representing the environmental consequences of human activities on a global scale. It was developed in the Netherlands by the RIVM (*Rijks Instituut voor Volksgezondheid en Milieu,* the Netherlands National Institute for Public Health and the Environment) and the MNP (*Milieu en Natuur Planningbureau*, the Dutch Environmental Assessment Agency) and is at the centre of intregrated models used by the IPCC and by the MA.

[3] IMPACT (International Model for Policy Analysis of Agricultural Commodities and Trade) is an equilibrium model of agricultural markets. This model, developed by the IFPRI, has been used for IPCC reports, those of World Water Vision in 2000, and those of the MA, as well as in the exercises of the International Assessment of Agriculture, Science and Technology for Development (IAASTD).

[4] The WaterGAP model is used to represent the impact of socio-economic scenarios in terms of water demand and therefore of pressure on water resources in the large catchment areas on a global scale. Developed by the CESR (Centre for Environmental Studies and Research, University of Kassel), it was used, in particular, in the framework of World Water Vision and then for the MA.

field to which they have not yet been applied, for example agriculture. This type of integrated modelling architecture is however highly complex and involves a large number of modelling teams. Any alteration to the initial scenarios would imply a new cycle of modelling. This explains why use has thus far been restricted to these four reference groups of scenarios. Within the framework of the International Assessment of Agricultural Knowledge, Science and Technology for Development (IAASTD), the panels have however challenged this choice of working from the four reference groups. They consider that the four groups are not suited to the analysis of issues concerning food and agricultural systems.

Agrimonde constitutes a complementary approach to these integrated assessment exercises. It makes it possible to more rapidly explore the coherence of the quantitative assumptions formulated in the scenarios, and is an incentive to focus a large part of the work on discussions on all the qualitative assumptions and on their coherence. This approach serves to build alternatives to the reference scenarios, and to explore development trajectories that are not currently represented in the four main groups of scenarios used in the early stages of integrated modelling. It could also enable emergent actors in the international debate on future food and agricultural systems (environmental NGOs, other non-profit organisations, etc.) to put forward for discussion new scenarios reflecting their perception of the challenges. In this respect, the Agrimonde tool is designed to enhance the capacity to generate coherent scenarios, before they can be modelled in various quantitative terms.

Chapter 2
Agribiom: A Tool for Scenario-Building and Hybrid Modelling

Bruno Dorin and Tristan Le Cotty

Agribiom is a quantitative tool devoted to the analysis of the world's production, trade and use of biomass. Its construction was initiated in early 2006 at CIRAD, with the aim of creating a tool for use in collective scenario-building, such as the Agrimonde project, and in hybrid modelling exercises[1,2].

At this stage, the (past and future) physical balance between food biomass resources and their use is the core issue and driver of Agribiom. Such balances can now be reconstructed (from the 1960s to date) or simulated on various geographical scales (from a country to the whole world) according to certain units and categories designed to:

- provide a tool for retrospective analysis and scenario-building that is sufficiently simple, all-encompassing and robust so that it can attract and mobilise a wide variety of expertise around questions of production, trade and consumption of biomass on national and global scales,
- collect and generate a set of data for developing new analyses and models, especially in fields and on scales in which statistical data and modelling exercises are limited (e.g. conversion of plant biomass into animal biomass on a national scale),
- characterise existing or potential modes of production and consumption of food biomass, and link the specificity of these modes to data, models or debates pertaining to food security, poverty, demand for non-food agricultural products

[1] Especially those undertaken with the CIRED (*Centre international de recherche sur l'environnement et le développement*) in cooperation with the CFE (*Conseil français de l'énergie*) on the subject of "Energy-food competition in land use" (Dorin et al. 2009; Dorin and Gitz 2008).

[2] Hybrid modelling consists in combining economic models with physical and technological data models (ed. note).

B. Dorin (✉) · T. L. Cotty
CIRAD—Umr CIRED, TA C56/15, 73 rue J.F. Breton,
34398 Montpellier Cedex 5, France
e-mail: bruno.dorin@cirad.fr

T. L. Cotty
e-mail: lecotty@cirad.fr

S. Paillard et al. (eds.), *Agrimonde – Scenarios and Challenges for Feeding the World in 2050*,
DOI 10.1007/978-94-017-8745-1_2, © Éditions Quæ, 2014

(biofuels, biomaterials, etc.), international trade, exploitation and prices of minerals or other natural resources, greenhouse gas emissions or sinks, conservation of services rendered by ecosystems, and so on.

To fulfil these functions, Agribiom is divided into four work packages, consisting in:

1. collecting, verifying and collating, over several decades, millions of data on national production, trade and uses of agricultural and food products,
2. using these basic data to generate new statistical series that serve for new analyses and new modelling exercises,
3. constructing an interface so that these data and models can easily be shown to various stakeholders (researchers, experts, policy-makers, entrepreneurs, NGOs, etc.), with a view to discussing them and then simulating and debating over various resource-use scenarios for food biomass,
4. interacting with other quantitative tools, especially computable general equilibrium[3] and biophysical models.

This chapter describes the progress made in the first three work packages. Chapter 3 shows some of Agribiom's outputs for a brief retrospective analysis of the world food economy (1961–2003), and in the following chapters, outputs for the Agrimonde scenarios are presented, along with related assumptions and interpretations.

General Organisation of Data Processing

To meet the objectives in terms of retrospective analysis, production of new statistics and models, simulations of new resource-use accounts of food biomass, interaction with various expert knowledge or numerical models, a huge number of historical data are fed into Agribiom (over 30 million non-redundant values in 2008). The treatment of this mass of information is illustrated in Fig. 2.1. SAS® and Microsoft Access® software is used to ensure:

- traceability of operations and calculations thanks to an arrangement of SAS programs between raw data files with variable structures and formats (xls, csv, txt, etc.) and the databases and models developed for the exercise,
- the convergence of these databases and models towards an interface constructed with Access (database management system with SQL code) to view and exploit these databases and models on various possible geographical scales (from a single country to the whole world).

[3] These are macroeconomic models (dealing with a whole economy) that include all activities, production factors and institutions, and therefore all markets (editor's note).

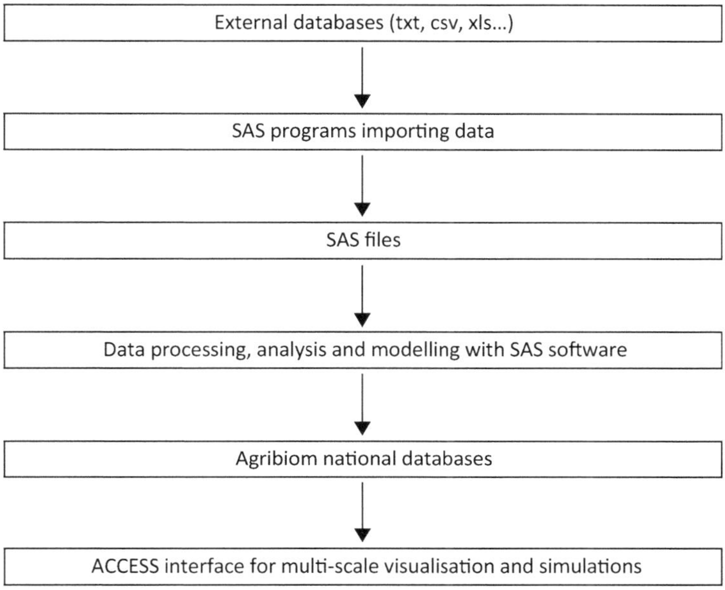

Fig. 2.1 General organisation of data processing

Temporal and Geographical Coverage

The United Nations FAOSTAT service collects, harmonises and disseminates a huge volume of national data on food and agriculture. This large quantity of data can be explained by the FAO's concern to include all agricultural products for human consumption (and not only those traded between countries) as well as all countries (and not only those with enough resources and skills to provide detailed good-quality statistics). Most of our work draws upon these FAOSTAT databases, even though they have certain shortcomings, largely related to the above-mentioned concern for exhaustiveness. These shortcomings can easily be highlighted, as well as the greater reliability of certain other databases focused on particular periods, products or regions. In our work, we favoured a "macroscopic" approach over a "microscopic" approach focused on specific fields, as we were keen, as far as possible, to obtain wide, all-encompassing views of a vast (geographical and historical) landscape rather than a few selective precise photographs of it. This focal distance for observing, analysing and modelling is complementary to others; it affords access to knowledge to which others do not have access, and vice-versa.

Improvements in the reliability and coherence of FAOSTAT data are nevertheless desirable, along with their expansion to areas in which there are no (or no longer) structured series harmonised on an international scale (especially concerning agricultural production factors). In this respect, in June 2006, when the Agrimonde project was launched, the FAO itself embarked on a vast and ambitious reform intended mainly to improve its Supply Utilization Accounts (SUA) and the data comprising them. To do so it excluded data series going back to 1961 ("$FAOSTAT_1$") and, in 2007, proposed new series starting in 1990 ("$FAOSTAT_2$"). Because of this closer historical focus, and for other more technical reasons (reorganisation of product lines and their coding, of the format of basic files, of the content of sections, etc.), we preferred to use the $FAOSTAT_1$ series (FAO 2006). However, the FAO abandoned $FAOSTAT_2$ at the beginning of 2008 and $FAOSTAT_1$ was resumed. This goes to explain why, in 2009, the FAO's SUA and Food Balance Sheets had not been updated beyond 2003. The same applies to most of our series.

Between 1961 and 2003, the earth's surface area did not change—unlike the number of countries and their borders. In the FAOSTAT series, over 250 geographical units have been recorded over the past four decades. We selected 149 units (Appendix 1, p. 241), after excluding a large number of islands and micro-states for which very little or highly irregular data were available, as well as some larger areas with the same lack of reliable statistics: Afghanistan, Antarctica, Bhutan, Iraq, Oman, Papua New Guinea, Western Sahara, and Somalia[4]. In 2000, in comparison with the total (excluding Antarctica) of the FAO for the same year (total named "World+"), this selection represents:

- 98.3 % of human populations (5,983,885 Minhab./6,085,574),
- 98.6 % of cultivated areas (food and non-food crops) (1,512,948 Mha/1,534,945),
- 97.3 % of land areas (13,078,385 Mha/13,443,345).

Our "World" total will therefore be calculated with the entities specified above, the number of which varies from one year to the next: e.g. after 1991, the entity "Soviet Union" was divided up into 15 new units (Russian Federation, Ukraine, etc.). The same applies to the regional totals of a particular zoning of the world. For the Agrimonde scenarios, the zoning used is that of the Millennium Ecosystem Assessment (MA) which groups together countries (or divides up the world) into six regions (Fig. 2.2) (MA 2005b). The distribution of our entities across the six MA regions is detailed in Appendix 1.

This zoning delimits regions considered to be relatively homogeneous according to some indicators. The choice of other indicators could have accounted for their very real internal ecological, socio-economic or historical diversity, with results varying according to the geographical units chosen to carry out the analysis (district, country, etc.). This grouping of areas and change of scale of analysis is necessary, even though it raises various important questions in the estimation of certain values, as in the development and application of models.

[4] The "Belgium-Luxemburg" zone was maintained, whereas from 2000 onwards, the series pertaining to food balances had no data for this set or for either of its units separately (Belgium or Luxemburg). This introduced a slight bias into several evaluations.

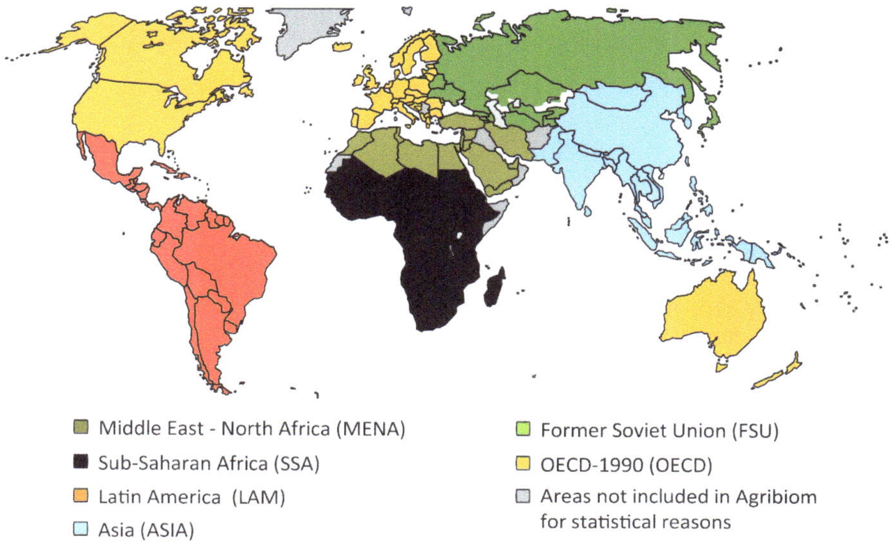

Fig. 2.2 Map of the six regions of the MA. (Source: based on MA (http://www.millenniumassessment.org/documents/document.774.aspx.pdf) Cartographic source: Articque)

Human Populations

Agribiom annual historical country data on human populations are drawn from the FAOSTAT "PopStat" series[5]. This series groups together two sets of estimates by the United Nations Population Division: the "Population-Estimates 2004 rev." and the "Population-Estimates 2006 rev.". The first set was chosen because it describes populations according to their gender (female/male), their dwelling place (rural/urban)[6], their dependence on agriculture ("agricultural population")[7] and their labour force ("economically active population")[8]. The "2006 rev." data set is less

[5] File called "PopSTAT-Annual-Time-Series1" in 2007–2008.

[6] Online FAOSTAT glossary (2008): "Rural population" = "Residual population after subtracting urban population from total population" and "Urban population" = "Usually the urban areas and hence the urban population are defined according to national census definitions which can be roughly divided into three major groups: classification of localities of a certain size as urban; classification of administrative centres of minor civil divisions as urban; and classification of centres of minor civil divisions on a chosen criterion which may include type of local government, number of inhabitants or proportion of population engaged in agriculture, as urban".

[7] Online FAOSTAT glossary (2008): "Agricultural population is defined as all persons depending for their livelihood on agriculture, hunting, fishing and forestry. It comprises all persons economically active in agriculture as well as their non-working dependents. It is not necessary that this referred population exclusively come from rural population".

[8] Online FAOSTAT glossary (2008): "The economically active population refers to the number of all employed and unemployed persons (including those seeking work for the first time). It covers

complete but more up-to-date (2006 for the "2006 rev." data set as opposed to 2005 for the "2004 rev." data set). These two sets give substantially different figures:

- at the beginning of the period (1960s). For 1961, the world population (FAO aggregate "World+") is respectively 3,081 and 2,804 billion,
- at the end of the period (2000s). For 2005, it amounts to 6,465 billion individuals in the first set and to 6,515 in the second.

For projections of human populations in 2050, two data sources per country were mobilised:

- MA projections according to four scenarios (Adapting Mosaic, Global Orchestration, Order from Strength, TechnoGarden), from 2000–2050 in 5-year segments and 21 age groups (0–4 years old, 5–9, etc.) (MA 2005b),
- United Nations projections published on line in 2007 on the UNSTATS website[9], for every year from 2006–2050, based on four hypotheses: a constant fertility scenario, a high variant projection, a low variant projection, and a medium variant projection.

Since our geographical area of study does not include all countries, for reasons outlined in the preceding section, and is not altered in retrospective and prospective analyses to ensure coherent calculations and unbiased comparisons over time, discrepancies are found with our "total world" and the "total world" from other sources, as shown in Table 2.1.

Land Use

Annual historical country data on general land use combine three series of FAOSTAT data:

- [1] the "Land" series as in 2007[10],
- [2] the "Land" series as in 2006[11],
- [3] the "Irrigated area" series as in 2006[12].

The series [1] updates FAO data on land use up to 2005, in which six categories are distinguished:

---

employers; self-employed workers; salaried employees; wage earners; unpaid workers assisting in a family, farm or business operation; members of producers' cooperatives; and members of the armed forces. The economically active population is also called the labour force".

9 "Total population (UN Pop. Div. annual estimates and projections) [code 13660]" downloaded on 08/05/2007 at http://unstats.un.org/unsd/cdb/cdb_advanced_data_extract.asp?srID=13660

10 "RessourceSTAT-Land1.xls" file.

11 "9541E_0.csv" file.

12 "9542E_0.csv" file.

Table 2.1 Variations in estimates of world human populations (2000 and 2050)

Year	Source	Total countries (million inhabitants)		Dif.	
		FAOSTAT, UNSTAT, MA	With Agribiom	(Million)	(%)
2000	FAOSTAT—Estimates 2006 Rev.	6,124	5,984	140	2.3
	FAOSTAT—Estimates 2004 Rev.	6,086	5,984	102	1.7
	UNSTAT, 2007—Code 13660	6,086	5,984	102	1.7
2050	MA, 2005—Scenario GO	8,085	7,872	213	2.6
	MA, 2005—Scenario TG	8,812	8,578	234	2.7
	MA, 2005—Scenario AM	9,514	9,265	250	2.6
	MA, 2005—Scenario OS	9,559	9,303	256	2.7
	UNSTAT, 2007—Low variant projection	7,667	7,440	227	3.0
	UNSTAT, 2007—Medium variant projection	9,060	8,800	260	2.9
	UNSTAT, 2007—High variant projection	10,627	10,330	297	2.8
	UNSTAT, 2007—Constant fertility scenario	11,634	11,245	389	3.3

Scenarios: *GO* Global Orchestration; *TG* TechnoGarden; *AM* Adaptating Mosaic; *OS* Order from Strength

- annual crops (called "Arable land")[13],
- plantations ("Permanent crops")[14],
- pastures ("Permanent meadows and pastures")[15],
- forests ("Forests and woodland")[16],
- other emerged land ("Other land"),
- lakes, rivers and other immersed land ("Inland water").

[13] Online FAOSTAT glossary (2008): "Arable land is the land under temporary agricultural crops (multiple-cropped areas are counted only once), temporary meadows for mowing or pasture, land under market and kitchen gardens and land temporarily fallow (less than five years). The abandoned land resulting from shifting cultivation is not included in this category. Data for 'Arable land' are not meant to indicate the amount of land that is potentially cultivable".

[14] Online FAOSTAT glossary (2008): "Permanent crops are sown or planted once, and then occupy the land for some years and need not be replanted after each annual harvest, such as cocoa, coffee and rubber. This category includes flowering shrubs, fruit trees, nut trees and vines, but excludes trees grown for wood or timber".

[15] Online FAOSTAT glossary (2008): "Permanent meadows and pastures is the land used permanently (five years or more) to grow herbaceous forage crops, either cultivated or growing wild (wild prairie or grazing land). The dividing line between this category and the category 'Forests and woodland' is rather indefinite, especially in the case of shrubs, savannah, etc., which may have been reported under either of these two categories".

[16] Online FAOSTAT glossary (2008) for "Forests and Woodland": "Land under natural or planted stands of trees, whether productive or not. This category includes land from which forests have been cleared but that will be reforested in the foreseeable future, but it excludes woodland or forest used only for recreation purposes. The question of shrub land, savannah, etc. raises the same problem as in the category 'Permanent meadows and pastures'".

Table 2.2 Variations in estimates of the earth's land use (2003)

Surface areas		Total countries (1000 ha)		Dif.	
		FAOSTAT (world+)	With Agribiom	(ha)	(%)
(1) Crops and plantations	Total	1,551,518	1,529,043	22,475	1.4
	annual crops (arable land)	1,413,002	1,392,951	20,052	1.4
	plantations (permanent crops)	138,516	136,093	2,423	1.7
	irrigated area (total area equipped for irrigation)	276,500	270,273	6,227	2.3
(2) Pastures (permanent meadows and pastures)		3,415,704	3,325,988	89,716	2.6
(3) Forests (forest)		3,966,660	3,904,776	61,883	1.6
(4) Other emerged land (other land)		4,078,908	3,891,722	187,186	4.6
(5) Lakes, rivers & other (inland water)		429,780	426,910	2,870	0.7
Total (1)+(2)+(3)+(4)		13,012,789	12,651,530	361,260	2.8
Total emerged area (land area)		13,013,621	12,651,530	362,091	2.8
Total (1)+(2)+(3)+(4)+(5)		13,442,569	13,078,440	364,130	4.6
Total area (country area)		13,443,401	13,078,440	364,961	2.7

The total of these areas should theoretically be equal to the "total area of the country" (Table 2.2), a total which the FAO provided with other intermediate aggregates such as "Arable and Permanent Crops" (which we called "cultivated area"), "Agricultural Area" (cultivated area + pastures), etc.

With the publication of the series [1], FAOSTAT adds new and important sections (e.g. "Fallow land", "Temporary meadows & pastures"). However these were seldom provided, or were insufficiently updated until 2005 for irrigated areas, which were imported from the former series [3] covering the period 1961–2003. The series [1] also proposes new estimates for forests, without going back further than 1990. Before then, data from the series [2] were imported, and the "Other land" category was adjusted so that the sum of the six land use categories did not exceed the total surface area of the country. Finally, this series [1] does not correct certain shortcomings, errors and inconsistencies on land use noted in previous years in this series[17]: certain corrections had to be made in order to smooth some series.

In Agribiom, annual and permanent croplands are merged into a single category, "cultivated land". For the simulations, the following are distinguished within this category: the "food cultivated area" (FCA) and the potentially large "non-food cultivated area" (NFCA: rubber, tobacco, fibres, eucalyptus, etc.). Until the 2000s, these NFCA were considered as nil even though this was not the case[18], mainly for the following reason: we had decided not to use the "harvested areas" per crop as provided in the FAO series "prodSTAT", because our attempts to relate these data

[17] The surface area of Spain decreased then increased between 1990 and 2003; the surface area of Greenland increased by 6.9 million hectares between 1996 and 1997, etc.

[18] FAO figures indicate that in 2003, on the global scale ("Word+"), the (gross) harvested areas in fibres, rubber and tobacco totalled slightly over 46 million hectares, which represents 3 % of the (net) cultivated area (1,552 Mha).

to the net cultivated land of series [1] too often led us to surprising or inconsistent results.

Finally, annual national data on coasts and maritime areas (dated from 1990–2000 depending on the case) were imported from a database developed by the University of Hamburg. This database, temporarily available on the university's website (www.fnu.zmaw.de), compiled data of various origins (World Resources Institute, CIA World Fact Book, Delft Hydraulics, Gallup and Sachs, etc.) including The Global Maritime Boundaries Database (GMBD) for continental shelves.

Potentially Cultivable Lands

At the end of the 1990s, the aim of the FAO and the International Institute for Applied System Analysis (IIASA), via Fischer et al. (2001, 2000, 2002), was to carry out a new evaluation of the world's potential agricultural production by means of recent breakthroughs in satellite imagery and GIS (Geographical Information System) techniques. This approach, called Global Agro-Ecological Zones (GAEZ), is based on the Agro-Ecological Zones (AEZ) which has been the subject of various studies over the past 20 years. The aim of the AEZ is to identify and characterise climatic zones, soils and lands suitable or not for agriculture.

The GAEZ method consists roughly in adjusting and combining data at a geographical scale which is far finer than the national one, i.e. grid-cells of a few square kilometres. These data belong to the following two sets: data on agro-ecological environments, on the one hand, and data on possible land utilisation, on the other.

Data on Agro-Ecological Environments

These data concern:

- the climate (CRU data/model at 30 min[19] latitude/longitude, with 1961–1990 mean values for the so-called "reference" climate, and IPCC [Intergovernmental Panel on Climate Change] data for the climate scenarios),
- the type of soil (FAO/UNESCO DSMW database on 2.2 million cells of 5 min latitude/longitude),
- the slopes (digital elevation model GTOPO30 at 30 arc-second latitude/longitude),
- the "current" land occupation (12 "aggregate" types of occupation drawn from GLCC maps at 30 arc-seconds latitude/longitude, based on satellite images probably taken in 1992/1993).

[19] 1 degree (60 min) at the equator is equivalent to about 111 km; 30 arc-seconds = 1 minute = 1,854 km at the equator.

Data on Possible Land Utilisation

462 land utilisation types (LUT)[20] were characterised in GAEZ by combining:

- 154 agricultural products including some fodder and pastures, or rather 27 species broken down into various crop-types attached to a climate zone: 8 cereals (83 crop-types: 4 for hibernating wheat, 12 for non-hibernating wheat, 13 for grain maize, 6 for silage, etc.), 3 tubers (8 crop-types: 4 potatoes, 1 manioc, 3 sweet-potatoes), 3 peas and lentils (17 crop-types), 6 oilseeds (25 crop-types, of which 1 palm and 1 olive), 1 fibre (7 crop-types for cotton), 2 sugar crops (6 crop-types: 1 sugar cane and 5 sugar beet), 1 fruit (banana/plantain) and 3 fodder (5 crop-types: 1 alfalfa, 4 pastures of forage grass, 4 pastures of leguminous fodder plants),
- 3 levels of input and management, successively labelled "low" (no use of chemical fertilisers, pesticides or improved seed), "intermediate" (use of certain "modern" inputs and partial mechanisation) and "high" (similar to commercial farming as practised in Western Europe and North America).

The combination of these data in each grid-cell led the GAEZ team first to calculate potential yields (of biomass and of product harvested) without any constraint other than temperature and solar radiation, and then to revise these yields successively in relation to:

- "agro-climatic constraints" (rainfall, mainly),
- "soil and terrain constraints", including, in particular, slopes (which restrict the intensification of production via mechanisation, irrigation, etc.) and the need to leave land fallow (to ensure long-term fertility of soil in the area under consideration).

These constraints were used by the team to estimate yields in the case of rainfed and irrigated crops (without assuming real availability of water nor the quality thereof), and according to the three levels of inputs and management mentioned above (low, intermediate and high).

The final outputs are estimates of surface areas (1,000 ha), by crop (wheat, rice, etc. with some aggregated categories, including "cereal crops" and "all crops"), by input level (low, intermediate and high) and by the use or otherwise of irrigation (at least for the high and intermediate input levels), for four "suitability classes" for agriculture: VS (very suitable), S (suitable), MS (moderately suitable) and mS

[20] The crop catalogue database provides a quantified description of LUT. Factors included are crop characteristics such as: duration of crop growth cycle, duration of individual crop development stages, photosynthetic pathway, crop adaptability group, maximum leaf area index, harvest index, development stage-specific, crop water requirement coefficients, yield reduction factors relating to moisture stress and yield loss, food content coefficients (energy, protein), extraction/conversion rates, crop by-product/residue coefficients, and commodity aggregation weights. Also included are parameters describing, for both rain-fed and irrigated LUT, thermal requirements, growing period requirements, and soil and terrain requirements, applicable in tropical, subtropical, temperate, and boreal environments, respectively.

(marginally suitable), along with NS (not suitable land) and NAG (land for settlement and infrastructure). The exercise also led to the estimation of maximum potential yields for each of the combinations listed above.

All the GAEZ data expressed per country and available on line (FAO and IIASA 2000) were imported, and those used for the Agrimonde project (after aggregation per MA region) are presented in Fig. A2.4. The examination of these data reveals certain difficulties:

- interpretation of "All crops" and "Mixed inputs" aggregates is difficult due to fuzzy definitions,
- the areas presented in the three sources (FAO and IIASA 2000; Fischer et al. 2000, 2002) are not identical[21],
- GAEZ country areas are much the same as those of the FAO (see "Land use", p. 30) however there are exceptions[22] which cannot be explained by immersed land areas,
- the year of evaluation for surface areas of forests and lands for settlement and infrastructure (housing, roads, etc.) is not provided (probably 1992/1993),
- GAEZ potentials are not expressed by category of current land use, except for forests. However total forest surface areas are very different (generally far inferior) to those of FAOSTAT and, more generally, to other sources of data on land use during the 1990s,
- similar estimates of potential croplands were made after simulation of different scenarios of (uniform) global warming (+1 °C, +2 °C, etc.), but these data have proved to be inaccessible.

Food Biomass Resource-Use Balances

As announced in the introduction to this chapter, the core subject of Agribiom is the balance—either reconstituted for the past or simulated for the future—between food biomass resources and their use, with three particularities.

The first particularity is that our resource-use balances are calculated for the near totality of "food biomass" that is divided into five "compartments" based on the origin of the products and on land use:

- vegetal/plant products (vege),
- animal products, divided into products of grazing animals comprising ruminants and large herbivorous animals (rumi), and products of non-grazing or monogastric animals (mono),

[21] Example: for the VS + S + MS potential in rainy conditions with a mixed level of inputs, we find successively, for North America, 384.2 Mha in online data, 405.5 Mha in the 2000 report, and 366.3 Mha in the 2002 report.

[22] Bhutan (14 % difference), Suriname, Liberia, Morocco, Ecuador, Belgium-Luxembourg, Saudi Arabia, United Arab Emirates, Libya, Netherlands, Kuwait, India, Rwanda, Niger, Guinea-Bissau, Tunisia (6 %), etc.

- aquatic products (plant or animal), divided into freshwater products (aqua) and marine products (mari).

By "food biomass" we mean any organic matter that, in its primary form, can either serve as food for human consumption—and that does effectively serve that purpose in a form that is processed to a greater or lesser degree (grain, oil, bread, cornflakes, etc.)—, or else that is entirely (e.g. grains of maize) or partially (e.g. oilcakes) used for animal consumption or other purposes (seed, ethanol or biodiesel, green chemistry, etc.). This definition therefore encompasses a wide range of agricultural products but not such products as rubber, plant fibres, silk, wool, leather, essential oils, fodder (alfalfa, silage, straw, bagasse, etc.), and so on[23].

The second particularity of our resource-use balances is that we use food calories (kcal) as a common unit of volume for consumption, production and trade of biomass. All food biomass provides energy for humans. This amount of energy, per gram or kilogram of product ingested, is particularly high with plant or animal oils and fats, and particularly low with fresh produce such as citrus fruit, tomatoes, shellfish, and tropical products such as tea, coffee or pineapples. This unit is used to obtain the sum of (and group into "compartments") quantities of products that cannot feasibly be added up when they are expressed in tons, litres or other units. Yet, even though the analysis of food calories has several advantages, it also has limitations, especially from two points of view: economic (the value of a calorie of a grain of maize is not equivalent to that of a grain of coffee) and nutritional (Deaton and Dreze 2009; Dorin 1999). In this respect, it is important to highlight here that a satisfactory diet as regards calorie content does not necessarily have the required micronutrients (vitamins and minerals, particularly present in fruit and vegetables) nor even macronutrients (carbohydrates, proteins and lipids), the diverse forms of which have to be consumed in the right quantities (neither too much nor too little) if a person is to live a healthy and active life. In view of these and other considerations, we tried to express our caloric balances as far as possible according to their carbohydrate, protein and lipid content, on the basis of an average content of, respectively, 4, 4 and 9 kcal per gram.

The third particularity is that the annual resources and uses of food biomass are represented and then simulated according to the structure of the equation presented below:

$$
\begin{aligned}
&(\mathrm{area}_{r,i}\mathrm{x}(\mathrm{prod}_{r,i}/\mathrm{area}_{r,i})) - \mathrm{trad}_{r,i} + \Delta\mathrm{stoc}_{r,i} = \\
&(\mathrm{popu}_{r}\mathrm{x}(\mathit{food}_{r,i}/\mathrm{popu}_{r})) + \mathrm{feed}_{r,i} + \mathrm{seed}_{r,i} + \mathrm{vana}_{r,i} + \mathrm{wast}_{r,i}
\end{aligned}
$$

where:

i is a compartment of food biomass (vege, rumi, mono, aqua, mari)

[23] Our resource-use balances do not include live animals (although their trade and stock variations, in particular, do have an impact on food balances). One of the reasons is that only their products (milk, meat, etc.) are taken into account in the SUA of the FAO.

r is a region of the world (country or group of countries: MENA, SSA, LAM, ASIA, FSU, OECD)

$area_{r,i}$ is an area (ha) in a region r: the food cultivated area when i=vege, the inland water area when i=aqua, the continental shelf area when i = mari; otherwise (i = rumi, mono): area=1

$prod_{r,i}$ is the production of foodstuffs i in a region r (kcal)

$prod_{r,i}/area_{r,i}$ is the yield of foodstuffs i (kcal/ha) in a region r when i=(vege, aqua, mari); otherwise (i = rumi, mono), $prod_{r,i}/area_{r,i} = prod_{r,i}$

$trad_{r,i}$ is the net trade balance (total exports—total imports) of foodstuffs i (kcal) in a region r

$\Delta stoc_{r,i}$ is the stock variation of foodstuffs i (kcal) in a region r (negative sign if destocking)

$popu_r$ is the human population (inhabitants) in a region r

$food_{r,i}$ is the quantity of foodstuffs i (kcal) used in a region r for feeding the human population, including wastage occurring in the household

$food_{r,i}/popu_r$ is the average food consumption (including wastage) per person (kcal/capita) of foodstuffs i in a region r

$feed_{r,i}$ is the quantity of foodstuffs i (kcal) used in a region r for feeding animals

$seed_{r,i}$ is the quantity of foodstuffs i (kcal) used in a region r for reproductive purposes (seed, eggs for hatching, etc.)

$vana_{r,i}$ is the quantity of foodstuffs i (kcal) used in a region r for non-food purposes: lubricants, energy, cosmetic, biomaterial, etc.

$wast_{r,i}$ is the wasted quantity of foodstuffs i (kcal) in a region r between the general available quantities (Production—Exports+Imports+/- Stocks) and their allocation to a specific use (food, feed, etc.); this does not include losses occurring before and during harvesting, and wastage occurring in the household.

In agribiom, this equation must be verified:

- at the level of each biomass compartment i (e.g. vege, rumi),
- on the scale of each region r considered (e.g. MA regions),
- in such a way that the sum of the net trade balances (trad*:* exports-imports) per compartment i is nil on a global scale.

The volumes of biomass are expressed in terms of food calories, which may be total calories but also calories only from carbohydrates, proteins or lipids.

The first term of the equation represents the resources: regional biomass production plus or minus the net trade balance and net stock variations. For plant and aquatic biomass, regional production is represented as a function of the production area (ha) and its (partial) productivity in food (kcal/ha)[24]. As this representation is not possible for terrestrial animal biomass (rumi and mono)[25], other formulations had to be used for the simulation of such production (see "Animal production models", p. 42).

[24] This representation of production is very simple but raises certain questions (area and yield) rather than others (for instance, it does not allow the analysis of the size of the farming population and its—partial—labour productivity).

[25] The production of this biomass cannot easily be linked to specific numbers of hectares.

The second term of the equation represents regional biomass uses, of which the human food consumption is represented as human populations (number of people) who consume varying quantities of food biomass per person (kcal/capita).

This representation of resources and uses of food biomass is closely related to the statistical series that could reasonably provide historical data, on the scale of each country in the world, and over a relatively long period. The series mobilised for human populations (popu) and areas (area) are presented above (see "Human populations" and "Land use" page 30). The others are derived from a far larger database which contains and compacts detailed data series on the production and trade of agricultural products: the Commodity Balances of the Supply Utilization Accounts (SUA) compiled by the FAO (FAO 2006). The SUA have the major advantage of being developed: 1) for almost all countries in the world; 2) for more than 40 years (1961–2003); 3) for over 120 product lines; 4) and so that, for each of these lines, the evaluation of national "availabilities" (production + imports—exports—stock variations) shows a balance with the evaluation of national "use". These uses are broken down into six sections: the five mentioned above (food[26], feed[27], seed[28], vana[29], wast[30]) and a sixth called "food manufacture" (cf. infra).

---

[26] Online FAOSTAT glossary (2008): food "Data refer to the total amount of the commodity available as human food during the reference period. Data include the commodity in question, as well as any commodity derived therefrom as a result of further processing. Food from maize, for example, comprises the amount of maize, maize meal and any other derived products available for human consumption. Food from milk relates to the amounts of milk as such, as well as the fresh milk equivalent of dairy products".

[27] Online FAOSTAT glossary (2008): feed "Data refer to the quantity of the commodity in question available for feeding the livestock and poultry during the reference period, whether domestically produced or imported".

[28] Online FAOSTAT glossary (2008): seed "Data include the amounts of the commodity in question set aside for sowing or planting (or generally for reproduction purposes, e.g. sugar cane planted, potatoes for seed, eggs for hatching and fish for bait, whether domestically produced or imported) during the reference period. Account is taken of double or successive sowing or planting whenever it occurs. The data of seed include also, when it is the case, the quantities necessary for sowing or planting the area relating to crops harvested green for fodder or for food (e.g. green peas, green beans, maize for forage). Data for seed element are stored in tonnes (t). Whenever official data were not available, seed figures have been estimated either as a percentage of supply (e.g. eggs for hatching) or by multiplying a seed rate with the area under the crop of the subsequent year".

[29] Online FAOSTAT glossary (2008): vana "Data refer to quantities of commodities used for non-food purposes, e.g. oil for soap. In order not to distort the picture of the national food pattern quantities of the commodity in question consumed mainly by tourists are included here (see also "Per capita supply"). In addition, this variable covers pet food".

[30] Online FAOSTAT glossary (2008): wast "Amount of the commodity in question lost through wastage (waste) during the year at all stages between the level at which production is recorded and the household, i.e. storage and transportation. Losses occurring before and during harvest are excluded. Waste from both edible and inedible parts of the commodity occurring in the household is also excluded. Quantities lost during the transformation of primary commodities into processed products are taken into account in the assessment of respective extraction/conversion rates. Distribution wastes tend to be considerable in countries with hot humid climate, difficult transportation and inadequate storage or processing facilities. This applies to the more perishable foodstuffs, and especially to those which have to be transported or stored for a long time in a tropical climate.

These annual country accounts are in tonnes. For the 109 lines of what we consider as "food biomass" (Appendix 1, p. 241), these tonnages have been converted into total calories and into calories derived from macronutrients (carbohydrates, proteins, lipids), based on FAO references (2003), sometimes USDA references (2006), and on the equation $\text{Kcal}_{\text{total}} = (4 \times g_{\text{carbohydrates}}) + (4 \times g_{\text{proteins}}) + (9 \times g_{\text{lipids}})$. In the particular case of feed (e.g. soybean oilcakes), calorie and macro-nutritional equivalents have been subtracted from the calorie and macro-nutritional values of the primary product (e.g. soybean), from the calorie and macro-nutritional values of a secondary product (e.g. soybean oil), and from a world average extraction rate of that product calculated with the FAO's SUA tonnages for the entire period under consideration (e.g. 18 % for soybean oil). Once these conversions into calories had been performed, the lines were aggregated into compartments, as shown in Appendix 1, with few specific cases subject to questionable allocation[31].

The SUA offers a unique data source for assessing and analysing general trends in production, trade and use of biomass. However this accounting is imperfect and complex. In particular, we had to formulate and test various options for classifying lines into "primary" or "secondary" products, in order to avoid double counts (especially for production) and finally to obtain relatively balanced resource-use ratios on a global scale, in terms of total calories as well as macronutrients, over 43 years, without the "food manufacture" section. This section relates to volumes of "primary" products (e.g. groundnuts, produced locally and/or imported), used for local production of one or several "secondary" products appearing in the SUA (e.g. groundnut oil and groundnut oilcakes) according to processing yields for which data are not available. These difficulties are compounded by the fact that some products such as alcohols are derived not from a single primary product but from several products (cereals, grapes, sugars), which may themselves be "secondary" products (sugars in particular, from sugar beet or sugar cane). After multiple tests on all the countries for the whole period 1961–2003, we treated, for example, sugars and molasses as primary products, and consequently excluded from the analysis the volumes of sugar cane and sugar beet from which they were obtained[32].

Furthermore, a perfect resource-use balance is not obtained because the export volumes do not match the import volumes. These problems, among others, triggered a FAOSTAT reform in 2006, which was subsequently abandoned in 2008. With the calorie balances calculated here, we find that in the vast majority of cases the total use is less than total resources. This discrepancy can be explained in various possible ways[33] and is significant in several countries, especially the US where about

Waste is often estimated as a fixed percentage of availability, the latter being defined as production plus imports plus stock withdrawals".

[31] For instance, the line "Animal fats (raw)" was allocated to rumi even though this line is most probably also an output of mono and mari products.

[32] This biomass is not traded much internationally, therefore no bias is introduced at this level. However it may, as in China, be used as feed and not only for producing sugar.

[33] We note the absence of SUA data for Belgium and Luxembourg (referenced in our base) from the year 2000 onwards, although this region is a net importer of plant calories (about 50 Gkcal/day since 1975); complete absence of SUA data for countries excluded from our base for this reason

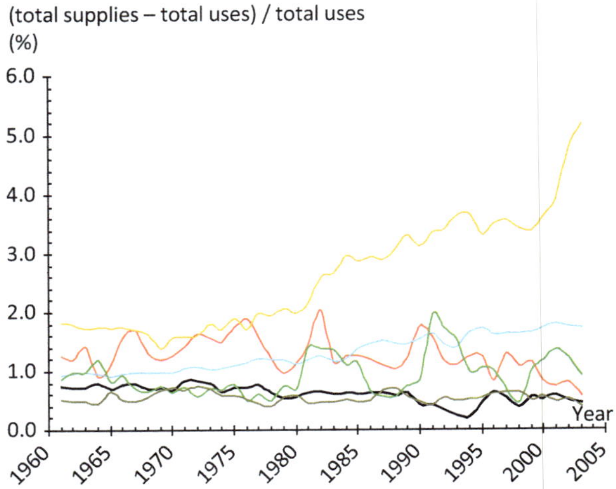

Fig. 2.3 Difference between supplies and uses of plant food calories (1961–2003)

10 % of plant food resources "disappeared" in this way in the early 2000s. But for the vast majority of countries, these gaps are far smaller: less than 3 % (US included) in the early 2000s on the world scale, and less than 2 % over 43 years for five of the six MA regions (Fig. 2.3). With stock variations, these gaps represent what we call "Residue" in the simulations.

Non-Food Biomass

The term "biomass" denotes a wide range of matter corresponding to differing conceptions and definitions, from organisms living underground or underwater, to the leaves of trees and birds in the sky, from organic matter in the process of formation to that which is fossilised in the form of oil, natural gas, coal, lignin or peat. Here, by "non-food biomass", we mean:

- the organic "by-products" or "residues" from harvests of "food biomass": straw, stalks and cobs, wool, leather,
- agricultural products (including from livestock farming and fishing) that cannot be consumed by humans in their primary form: rubber, cotton or other fibres, silk, alfalfa and other fodder, grass,
- the organic "by-products" or "residues" from harvests mentioned above,
- wood in various forms (trees, fuel wood, etc.).

(e.g. Iraq, Afghanistan, Somalia) whereas they are probably net importers of substantial quantities of food; under-estimation of certain uses (including waste); overestimation of production or exports; incorrect assumptions in our treatment of the section "Food manufacture"; etc.

Various series of FAO national data (SUA and other) make it possible to evaluate the tonnage of numerous types of non-food biomass listed above, along with the evolution of these volumes over recent decades, either directly (tobacco, rubber, fibres, wool, leather, fuel wood, industrial wood, etc.) or indirectly (crop residues, standing forest biomass, etc.). Even though compilation and processing of these data within Agribiom need to be continued, our simulation tool makes it possible to investigate the question of competition/complementarity between food and non-food biomass through:

- land use, with varying surfaces of forests, pastures and "non-food cultivated areas" (nfca),
- non-food uses of food products (vana),
- models of animal production using, among other things, quantities of food products used as feed (see feed and models below).

Animal Production Models

This section first shows the importance and the difficulty, in a resource-use approach like Agribiom, of estimating the links between animal productions and plant resources available for these productions. It then proposes a first system of estimation that can serve to capture significant differences between technologies existing in this field at the world level.

Problematic Data and Representations

Animal husbandry—here, of land animals only—provides food for human beings (milk and dairy products, meat, eggs, etc.), of which people tend to eat more when their income rises. Along with the growth of human populations, the demand for animal products is expected to increase steadily in the future. Animal husbandry also provides many other services, for instance for savings, transport and traction, fertilisation of the land (animal manure), cooking (dried dung), lighting, washing or cosmetics (tallow and other animal fats), clothing (wool, leather, feathers, down, etc.), the recycling of organic waste, the maintenance of landscapes and areas rich in carbon and biodiversity, etc. Animals also fulfil religious or social functions (e.g. pets). They directly and indirectly employ a large number of people, and use just over 80 % of so-called "agricultural" land, with 3.3 billion hectares of pastures[34] and over half a billion hectares of cultivated land[35]. Animal husbandry is also cause for

[34] See definition of "land use" p. 30 and Chap. 3, p. 58.

[35] According to our estimations based on FAO data, in 2000, about one third of plant calories consumed in the world were used for animal feed, with major variations in this rate from one region to another (see Chap. 3, p. 63).

concern when it comes to sanitary problems (epizootics) and environmental issues, especially regarding soil (erosion due to overgrazing), water (consumption, pollution) and greenhouse gas emissions (Steinfeld et al. 2006).

Yet, despite the importance of animal husbandry from an economic or ecological point of view, there is a serious lack of statistical data on the subject on a global scale. Animal products other than food are often poorly evaluated (and sometimes not at all), as are production factors other than "concentrates" (cereals and oilcakes): labour, capital, inputs such as veterinary products, etc. From the point of view of animal feed for instance, this is particularly problematic when representing the process of conversion of plant biomass into various kinds of animal biomass, for the purpose of global foresight related to land use. In particular, in the case of large herbivorous animals and ruminants, biomass other than concentrates can be (and is in fact) provided as a supplement or substitute: annual fodder[36], grasses (green or dried) and other types of biomass from meadows, pastures, savannah and various other areas (including forests), crop residues (straw, stalks, haulm, etc.), food residues (peels and other discarded parts), etc. Some authors have attempted to quantify these different animal food resources on national or continental scales, for example Devendra and Sevilla (2002), Wirsenius (2003), Bouwman et al. (2005) and Smeets et al. (2007). Along with quantities, the evaluation of the quality of this biomass is equally important but also poorly known (dry matter, digestibility, energy, proteins, etc.). Finally, for all these sources of feed, much like the others (concentrates), there is a third significant lack of statistics at national scales: the distribution of animal consumption of biomass per species (horses, cattle, sheep, pigs, poultry, etc.) and/or by animal product (milk, meat, etc.).

Agricultural and food foresight exercises however use rates of conversion A (a_{11}, ..., a_{nk}) of a particular biomass i (i_1, ..., i_n) into an animal product P (p_1, ..., p_k). The biomass i is generally limited to volumes of concentrates (cereals, oilcakes), and the products P to volumes of milk (cow, buffalo, goat, etc.), meat (beef, pork, etc.) and eggs, or to a type of animal (calf, cow, bull, etc.). Rates A depend on the units of volume used for i, which may be kilograms of dry matter (Bouwman et al. 2005, Delgado et al. 1999), kilograms of protein (Sebillote 2001), kilocalories (Collomb 1999; Griffon 2006; Malassis and Padilla 1986), etc. These rates A are evaluated in two main ways referred to as the "physiological approach" and the "statistical approach".

The "physiological approach" seeks to evaluate rates A in relation to animals' individual physiological needs (for their maintenance, nutrition, growth, lactation, draft power, pregnancy, etc.), to the composition of herds and flocks (breeds, age, sex and weight of animals), and to local characteristics of available biomass i. In concrete terms, this approach requires a large number of assumptions to be made

[36] In the SUA (Commodity Balances) of the FAO (2006), there are 5 lines for fodder: *Alfalfa for forage and silage*, *Clover for forage and silage*, *Maize for forage and silage*, *Rye grass for forage & silage* and *Sorghum for forage and silage*. These lines are rather limited in number compared to numerous other productions of fodder existing around the world. They are also not provided for large countries such as Brazil, China and India. Because of these limitations we chose not to use these FAO data on fodder, despite their importance.

when we work on national scales, for the past and, even more so, for the future. The "statistical approach" consists in evaluating A in relation to volumes i and P observed at a certain point in time in a certain area and, for the future, in maintaining or altering A according to various experts' assumptions, to be made on all future feed sources i (i_1, ..., i_n) and other production factors, as well as on the impact of these assumptions on each A value (a_{11}, ..., a_{nk}).

In both cases (physiological approach and statistical approach), the representations and coefficients used to simulate the future tend, in practice, to move closer to the situations that are better referenced today, such as industrial breeding and experimental stations aimed at improving the productivity of dairy, meat or egg farms. Even though major progress has been made and will continue to be made in industrial forms of production, we cannot outright exclude, in scenario-building exercises, the fact that other forms of livestock farming will still exist, will be improved or will emerge in the future. These may effectively exploit certain local resources, or provide various forms of income and services to agrosystems and populations with little financial and logistic capital, as in most countries of the South today.

In view of all these considerations, we attempted to improve the representation and modelling of animal food productions at global level. Our approach was resolutely statistical and was divided into two main phases:

- the first involved building a database connecting various national data: 1) relative to animal production and to agricultural production factors; 2) with a large number of countries and over a large number of years (1961–2003) in order to obtain a satisfactory sample of measurements reflecting varied technological options/evolutions; 3) using aggregates and units liable to reveal general and robust phenomena (vege, rumi, mono, aqua and mari compartments quantified in terms of food calories, proteins or lipids),
- the second involved searching, in this database—that we would have preferred to be more complete (on annual fodder consumption, crop residues, pasture quality, etc.)—, for the statistical relations between animal food production and variables liable to explain this production. This research was geared towards the elaboration of "animal production functions".

Following the first stage, it was shown, in particular, that the partial productivity of plant feed (cereals and oilcakes, mostly) was effectively highly variable in space and time, in terms of total calories (Fig. 2.4) or proteins (Fig. 2.5). Long-term simulations of animal production, with a fixed coefficient for this production factor only (cereals and oilcakes, mostly), therefore present limits that the second stage (animal production functions) aims to transcend.

Regional Animal Production Functions

In microeconomics, a production function expresses the relationship between the inputs used by a firm and its production. It indicates, in the form of an equation or graph, what the firm can produce, based on various quantities and combinations of

Fig. 2.4 Plant food calories used to produce one calorie of animal foodstuff (1961–2003)

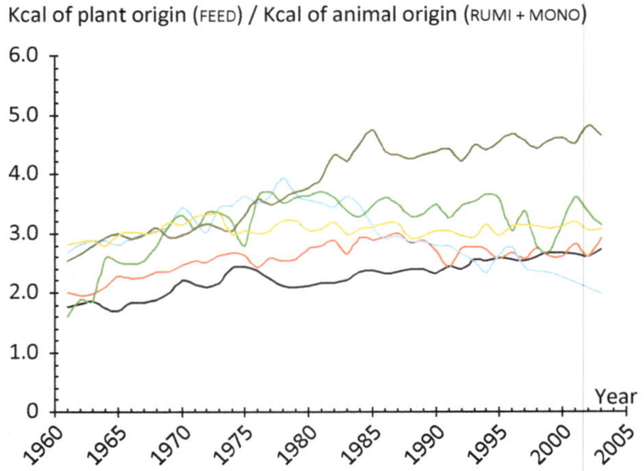

Fig. 2.5 Plant food proteins used to produce one protein of animal foodstuff (1961–2003)

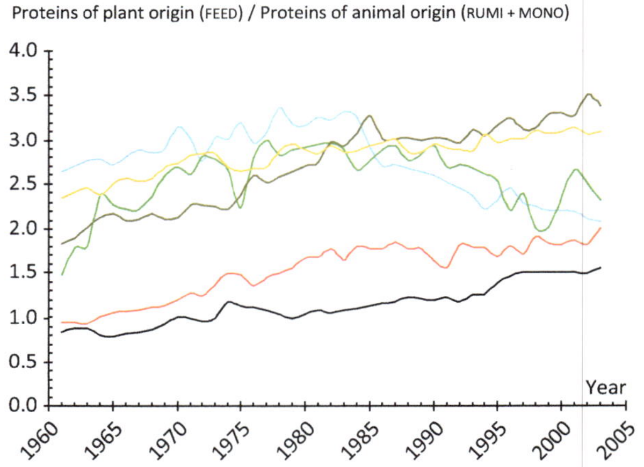

production factors. In general, the production function can be written $Q=f(x_1, \ldots, x_n)$ where Q is the quantity of an output, and $x_1, \ldots, x_n$ are quantities of production factors (labour, capital, inputs, etc.). This function can take different forms (linear, quadratic, Cobb-Douglas, CES, etc.), depending on the technology (whether marginal returns are decreasing or not, whether there are economies of scale or not, whether production factors are highly substitutable or not, etc.). This form is selected depending on the data and the aspects of the technology examined.

In our work, we sought to establish production functions:

- on the scale not of a firm but of a country (or of several countries grouped together in a region), which is often referred to as "cross-country production functions",
- using panel data over a 43-year period (1961–2003),
- in order to estimate annual productions of animal foodstuffs (milk, meat, eggs, etc.) converted into calories or protein equivalents (Gkcal) and grouped together

in two categories only: foodstuffs from "ruminants and large herbivorous animals" (rumi) and foodstuffs from "monogastric animals" (mono),
- with available data on some production factors used—entirely or partially—for these animal productions: feed of plant or animal origin (Gkcal of total calories or calories provided by proteins only), pasture area (thousands of hectares), human labour (thousands of farm workers), tractors (units), etc.

In other words, we sought to design multi-product production functions whose general form is $F(X, Y)=0$, where $X=(x_1, \ldots, x_n)$ is the vector of production factors and $Y=(y_1, \ldots, y_m)$ the vector of outputs produced with these factors. This type of function makes it possible, in particular, to distinguish the productivity of the feed in terms of products of rumi on the one hand and products of mono on the other. However, these functions are more difficult to estimate than mono-product functions when the allocation of factors to products is unknown, as is the case here (we know for example the total quantities of feed used in a country, but not those used respectively for rumi and mono). The allocation must therefore be derived from the aggregated estimates, by means of the various available methods (Just et al. 1983; Mishra 2007).

The estimation of such production functions also entails serious risks of biases that are identified in the literature, especially linked to the endogeneity of production factors. The correction of these biases requires appropriate estimation methods. Three estimation methods were selected:

- an autoregressive model that is an effective tool for eliminating autocorrelation (the error term of year t is used as an explanatory variable of year $t+1$),
- a generalised least squares estimation (weighted least squares and two-stage least squares) which substantially reduces the heteroskedasticity bias and, in most cases, gives estimation results close to the autoregressive model,
- different models with fixed effects, which potentially also help to correct endogeneity biases.

This led us to estimate and test various production functions:

- with a variable number of factors $(x_1, \ldots, x_n)$, and/or of composite indicators combining these factors with other available variables (to account for the quality of pastures in particular),
- with outputs and inputs expressed in the same units, either in total calories or in protein calories[37], especially to capture the "oilcake" effect (soybean cake in particular) which has increasingly become a protein supplement in feeding practices,
- with or without "*trend*" (to assess annual "technical progress"[38]) or temporal and geographical "*dummies*" (to capture the specific effects of certain years or countries),
- with the objective of modelling "geographical" production functions (for instance one function for each MA region) or "typical" production functions (e.g. "intensive-industrial", "extensive-agricultural", etc.)
- with different functional forms, especially linear and quadratic.

[37] Reminder: 1 g of proteins provides 4 kcal on average.

[38] Annual production increase not explained by the production factors of the production function.

For the Agrimonde foresight, the following properties were chosen:

- linear functional form,
- geographic functions (one for each of the six MA regions, including 12–40 countries per region),
- with neither *trend* nor *dummies,*
- using protein calories as a working unit; for the simulations, the conversion rates from protein calories into total calories are set equal to the last values observed (2003) but can be modified according to the scenarios (e.g. increase or decrease in the protein content of the feed),
- based on a system of two equations (production of proteins from rumi on the one hand and from mono on the other), with three explanatory factors: proteins from feed (plant and animal origin), hectares of pasture, and level of production of the "substitute" output (production of mono in production functions of rumi, and vice-versa).

These functions make it possible to fairly accurately reproduce the evolution of regional animal production over the past 40 years (Fig. A1.1). More elaborated functions can reproduce these past trends even more accurately, but this was not the major objective here. For the Agrimonde scenarios, the aim was to obtain functions which required a limited number of assumptions to be formulated for the simulations (in Agrimonde, each assumption is subject to time-consuming collective debates), and which tolerated a wide range of variation for the values of production factors (Agrimonde is a scenario-building exercise which can imagine very different worlds from those observed in the past).

The linear form is rather restrictive but is supported by a number of motivations:

- of all the forms tested, it is the most stable in the face of changes in geographical scales[39]; once the production function is estimated with national data, the coefficients of marginal productivity of each factor are valid for the countries of a region and for the entire region,
- the estimated coefficients are closer to physiological coefficients; for example, a coefficient of 0.2 associated with feed (in calories) means that one additional calorie of feed produces 0.2 additional calories of animal product (ruminant or monogastric), which represents a marginal conversion rate of 5 calories of feed per animal calorie; this coefficient is called "marginal productivity"[40],
- the linear form is compatible with a decreasing average productivity of the feed as it is observed empirically (Fig. 2.6). It is also compatible with a substitution between factors and with a substitution between outputs (at a fixed rate).

The generic form of the functions used for the Agrimonde scenarios is presented below (Tables 2.3 and 2.4), as well as their generalised least squares estimation (Tables 2.5 and 2.6). For the simulations, after setting regional quantities of feed and pastures, we solve a system of two equations and two unknowns for each region

[39] Including the Cobb Douglas function with constant returns to scale.

[40] A constant marginal productivity (as the linear form imposes) is a restriction since it does not allow for second order effects to be represented. On the other hand, it makes the model more robust.

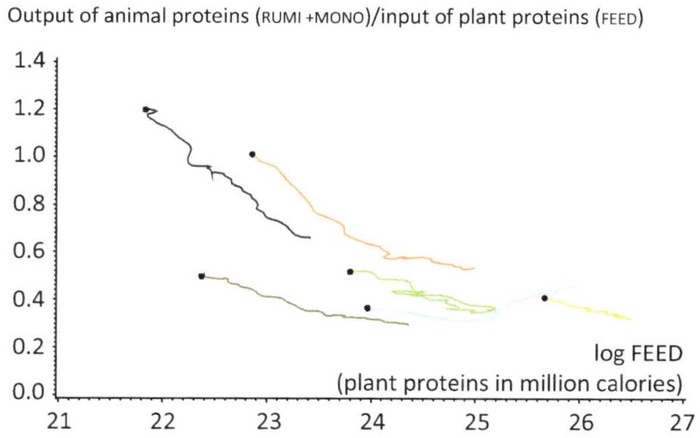

Fig. 2.6 Decreasing average productivity of plant food proteins in the production of animal food proteins (1961–2003). When the value of plant feed is low, the production of animal foodstuffs (*rumi* + *mono*) relies essentially on pastures, fodders or residues, and the average productivity of feed is then high (i.e. the Output/Input ratio is high). As the relative share of feed in the production increases, its average productivity decreases (the Output/Input ratio decreases). This evolution is generally accompanied by a decreasing share of rumi outputs in the total animal outputs

(prod_rumi and prod_mono). The result of this procedure respects the constraints of the two production functions, but excludes any possibility to fix in advance the rumi/mono proportion in the total amount of outputs. This possibility requires an alternative resolution that also exists in Agribiom, by choosing one or the other of the two production functions.

Interactive Interface and Simulations

One of the main objectives of Agribiom is to facilitate collective debates on past and future production, trade and use of biomass on a global scale, and to promote the emergence of common visions or questions on the past and the future. In order for it to become such a "mediating" tool, a great deal of time and care were devoted to the creation of an interface with Microsoft Access©. By the end of 2008, for various possible scales of geographical analysis (including the six MA regions), this interface was able to:

- show (through graphs) the 1961–2003 evolution of numerous variables obtained from the processing of several million historical data (Chap. 3, Figs. 3.1–3.13, Appendix 2, Fig. A2.1–A2.9), especially the variables or "parameters" which served to simulate the production, trade and use of food biomass,
- describe and test models devised internally (currently, animal production functions), by comparing their results to those observed in the past (1961–2003), by readily changing their coefficients (especially marginal productivities) or mode

Table 2.3 Generic cross-country animal production functions used in Agrimonde

$$prod\_rumi_k = \alpha_0^{\,j} + \alpha_1^{\,j} feed + \alpha_2^{\,j} pastures_k + \beta^j prod\_mono_k$$

$$prod\_mono_k = \gamma_0^{\,j} + \gamma_1^{\,j} feed_k + \delta^j prod\_rumi$$

Where:

- k is the country index
- J is the region index
- $prod\_rumi_k$ stands for the production of food proteins of ruminants, expressed in kcal per year, for country k
- $prod\_mono_k$ stands for the production of food proteins of monogastric animals, expressed in kcal per year, for country k
- $\alpha_0^{\,J}$ is the constant term (for all countries in region J) of the production function for rumi
- $\alpha_1^{\,J}$ is the marginal productivity of feed (of both animal and plant origin) in region J, expressed in kcal of proteins of rumi outputs (prod_rumi) per kcal of proteins of feed
- $feed_k$ is the feed use in country k, expressed in kcal of proteins
- $\alpha_2^{\,J}$ is the marginal productivity of pastures in region J, expressed in kcal of animal proteins per hectare of pasture area
- $pastures_k$ is the surface of the pasture area in country k
- β^J is the substitution coefficient between mono and rumi productions in region J
- $\gamma_0^{\,J}$ is the constant term (for all countries in region J) of the production function for Mono
- $\gamma_1^{\,J}$ is the marginal productivity of feed (of both animal and plant origin) in region J, expressed in kcal of proteins of mono outputs (prod_mono) per kcal of proteins of feed
- δ^J is the substitution coefficient between mono and rumi productions in region J

Table 2.4 Generic regional animal production functions used in Agrimonde

$$prod\_rumi_j = \sum_{k\in J} \alpha_0^{\,j} + \alpha_1^{\,j} \sum_{k\in J} feed_k + \alpha_2^{\,j} . \sum_{k\in J} pastures_k + \beta^j \sum_{k\in J} prod\_mono_k$$

$$prod\_mono_j = \sum_{k\in J} \gamma_0^{\,j} + \gamma_1^{\,j} \sum_{k\in J} feed_k + \delta^j \sum_{k\in J} prod\_rumi_k$$

where:

- $prod\_rumi_J$ is the production of food proteins of ruminants, expressed in kcal per year, for region J
- $prod\_mono_J$ is the production of food proteins of monogastric animals, expressed in kcal per year, for region J

of resolution, and by instantly visualising their results with new data or assumptions,

- enter, for a particular scenario envisaged (or one of its variants), the assumptions of parameters and models, and then debate, rework and finalise these assumptions by collectively simulating, with the interface, a global physical balance of the production, uses and trading of food biomass,
- archive the quantitative results obtained with the related assumptions, especially to make them transparent to other parties, and to allow for criticism or more in-depth development of the scenarios and their related assumptions.

Table 2.5 Estimates of cross-country production functions used in Agrimonde: rumi

	OECD	ASIA	LAM	SSA	FSU	MENA
$\alpha_0{}^J$ (Intercept)	+251,075,226.1	+20,528,324.9	+5,903,806.83	−2,668,975.75	+42,720,057.2	+19,928,565.5
$\alpha_1{}^J$ (feed)	+0.3 9***	+0.17***	+0.25***	+0.34***	+0.26***	+0.24***
$\alpha_2{}^J$ (pastures)	+6,411***	+3,000***	+12,881***	+3,349***	+2,758***	+4,081***
β^J (prod_mono)	− 1.89***	−0.01***	− 0.23***	−0.05***	−0.01***	−0.87***

*** significant at the 1 % level

Table 2.6 Estimates of cross-country production functions used in Agrimonde: mono

	OECD	ASIA	LAM	SSA	FSU	MENA
$\gamma_0{}^J$ (Intercept)	−16,343,932.5	−295,377,986	+1,749,430.94	+1,001,663.42	−368,017.209	+12,688,321.3
$\gamma_1{}^J$ (feed)	+0.15 ***	+0.39***	+0.21***	+0.18***	+0.08***	+0.16***
δ^J (prod_rumi)	−0.14 ***	−0.53***	−0.05***	−0.05***	−0.01***	−0.29***

*** significant at the 1 % level

The interface is organised into several windows or "parameterisation domains" (human populations, food consumption, land occupation, food production and productivity, food trade, food uses, animal production models, etc.) which make it possible to visualise historical data in each of the domains concerned, and then to register (or calculate), in each of these domains, values of scenarios (or variants of scenarios) on a specific timeline. A particular window can be used to:

- recapitulate, for each region of the world under consideration (here, MA zoning) and on the selected timeline, the main assumptions formulated for the scenario (populations, diets, land use, etc.) and their implications in terms of use, production and net trade (in Gkcal/day) for the five food biomass compartments (vege, aqua, mari, rumi, mono),
- adjust these assumptions until a physical balance between the uses and resources of food biomass is obtained on a global scale, some of these adjustments requiring the use of other Agribiom tools in order to be carried out correctly, especially as regards animal production[41].

A simulation via the Agribiom interface consists in illustrating a balance (or an imbalance) between the uses and resources of food biomass, considered by region and then globally. For each region considered, this illustration implies a specification of assumptions: 1) on the elements of our resource-use equation (see food biomass resource-use balances p. 36)[42], 2) on the models used to provide some of these elements (here, animal productions), and 3) on international trade, especially on the regional preferences for acquiring resources abroad (is there a preference for importing animal feed or animal products themselves? Which region could preferably supply them? etc.). When these assumptions are not all compatible, or to simulate the impact of a modification to one of them, adjustment criteria must be defined to select those variables which will be adjusted and those which will not.

In an economic equilibrium model such as IMPACT (IFPRI, International Food Policy Research Institute), the rules of adjustment are explicit and exogenous. The authors know them before carrying out a simulation. They are defined by a set of elasticities and constraints on certain physical or economic variables which lead to supply and demand functions. On the other hand, the quantities (production, consumption, surfaces, etc.) and equilibrium prices are generally endogenous. A difficulty often mentioned in these models probably stems from the choice of elasticities, that is, parameters which represent agents' reactions to

[41] Note that in its 2008 version, the Agribiom interface does not yet allow the assumptions and physical balances obtained to be associated with certain evaluations pertaining, in particular, to energy or water consumption, employment in agriculture, greenhouse gas emission or sink, etc. This was initially, and is still is, an objective.

[42] Except for stock variations which, for the simulation of the base year (e.g. 2003) chosen to serve as a reference for the study of other simulations, are integrated into a use section called "Residue". This "Residue" section also enables us to integrate amounts linked to statistical errors or inaccuracies found in the past (see food biomass resource-use balances); amounts without which there would not be a perfect equilibrium between resources and uses, and without which the comparisons of simulations then carried out would be biased.

variations in the economic environment (by how much does the wheat supply increase in a particular region when the wheat price increases by 1 %; by how much does consumption decline if the price in the region increases by 1 %; by how much does wheat consumption increase when the income of the region rises by 1 %, etc.). These elasticities are expected to provide stereotypical reactions of production and consumption to price variations or to variations of price-like economic variables (especially income). They have the important quality of making it possible to simultaneously implement many decentralised adjustments, while maintaining an economic equilibrium between supply and demand. Thus, an unsatisfied demand would be translated endogenously by a price increase that would trigger both an increase in production and a drop in consumption. The equilibrium between supply and demand is thus constantly guaranteed by price adjustments, and the (solvent) demand is always satisfied, by construction. But this category of models is not suited to representing a world geared towards the satisfaction of needs (physiological, social, environmental necessities, etc.); in this type of modelling, it is not for example certain people's lack of food nutrients (non-satisfaction of a need) that increases production, but the non- satisfaction of their demand, which depends on their purchasing power, preferences, and information.

In the Agribiom simulations, the quantities and other physical values are exogenous (i.e. chosen by a person or an expert panel). With each set of assumptions, we find a certain disequilibrium, with its distribution by region and biomass compartment. This disequilibrium is the endogenous (and relevant) information from the simulation. Based on this disequilibrium, there is an infinite number of ways of making an adjustment since, in practical terms, each element of the choice is continuous. From this point of view, the path proposed by the panel for reaching a balance, consisting of the alteration of certain elements by trial and error, following certain rules[43], defines a set of adequate conditions to obtain a resource-use equilibrium. It may also be useful, for the analysis, not to automatically balance the economy in order to show regional surpluses and deficits, and collectively to debate the different ways or conditions for remedying the deficits. One can also debate how needs can be met by simulating extremes: for example, an increase in demand can trigger an increase in yields without an increase in the cultivated area, or vice-versa. Extreme answers are probably not the most realistic, but they can be very valuable in a scenario-building exercise.

Thus, the added value of the Agribiom interface resides in learning the role of all the variables, models and decision-making rules used to achieve a global balance, and not only in the final image of the resource-use balance proposed at the end of the process. It is in this sense that the interface is interactive, and that it can only really function through interaction.

[43] Example of rules: (1) if a region faces a shortage in food calories, it imports the plant products necessary to cover the food needs of both humans and animals (i.e. domestic production of animal products with some imports of plant feed, instead of direct imports of animal products); (2) the imports come from the largest surplus regions, in decreasing order of their surplus quantities; (3) if total regional surpluses cannot cover total regional shortages at the global level, some exogenous variables are adjusted upwards (yields, cultivated areas, etc.) and not downwards (diet etc.).

Chapter 3
The World Food Economy: A Retrospective Overview

Bruno Dorin

This chapter provides a retrospective overview of the world foodeconomy over four decades (1961–2003), with a world divided into six regions [those of the Millennium Ecosystem Assessment (MA)], and using material generated by Agribiom. We will outline here how Agribiom can be used to analyse or revisit the past, while subsequent chapters will show how it can be used to debate and simulate new possible states of the world, along with all the assumptions and interpretations associated with these simulations.

This historical material was made available to the Agrimonde foresight panel to fuel its analysis and discussions. It inevitably oriented its world view, on the past as well as the future. The data and figures shown in this chapter—among a large amount of other Agribiom material—are broken down into six topics: human populations, food consumption, land use, food production and productivity, use of food products and, finally, food trade. In the scenario-building phase of Agrimonde, assumptions were made on these topics, based on past trends and on each scenario under consideration for 2050.

Human Populations

The world's human population has more than doubled over the past four decades: from 3 billion people in 1961 to nearly 6.2 billion in 2003. This is equivalent to a mean annual growth rate of 1.7%. Over half of this population (54% in 2003) is concentrated in ASIA, even though the population growth of this region (1.9% per annum) has not been the highest in the world [2.6% and 2.7% in Middle East—North Africa (MENA) and sub-Saharan Africa (SSA) respectively] (Fig. 3.1). ASIA also has the majority of the world's farmers: three-quarters of the world's economically active agricultural population in the early 2000s (Fig. 3.2), i.e. over a billion

B. Dorin (✉)
CIRAD—Umr CIRED, TA C56/15, 73 rue J.F. Breton,
34398 Montpellier Cedex 5, France
e-mail: bruno.dorin@cirad.fr

S. Paillard et al. (eds.), *Agrimonde – Scenarios and Challenges for Feeding the World in 2050*,
DOI 10.1007/978-94-017-8745-1_3, © Éditions Quæ, 2014

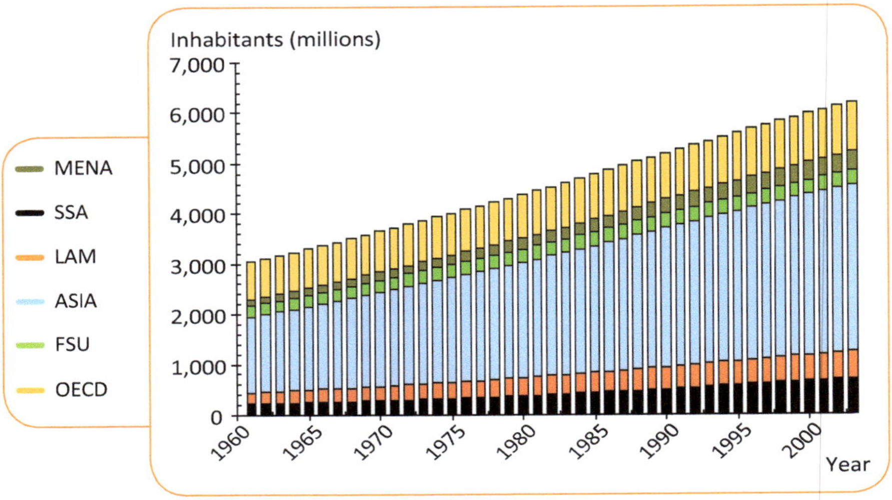

Fig. 3.1 Human populations (1961–2003)

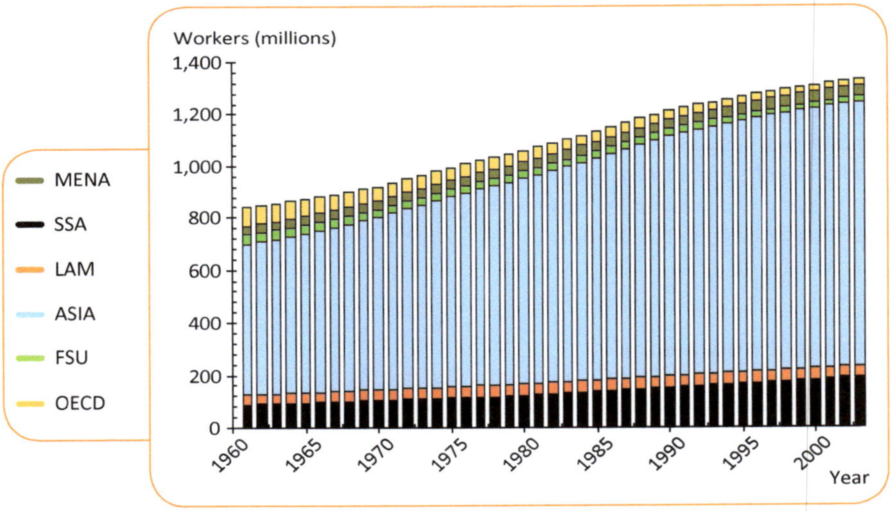

Fig. 3.2 Agricultural working population (1961–2003)

individuals living primarily in rural areas with their families and other workers directly dependent on the sector (traders, agricultural input suppliers, food processors, etc.). This agricultural working population has constantly grown in this region, as in most of the other regions [LAM, MENA and SSA], but at a far slower rate than that of the urbanisation of populations (Appendix 2, Fig. A2.1). In 2003

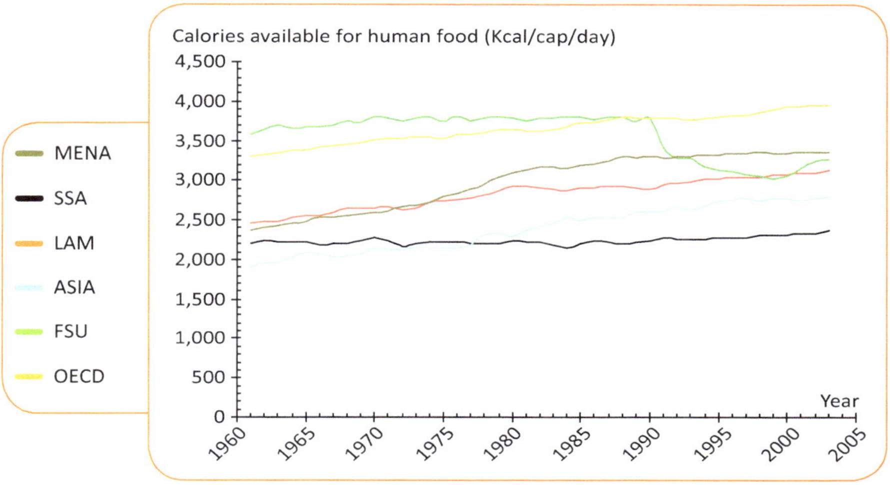

Fig. 3.3 Total calories available for human food (1961–2003)

half of the world's population was living in urban areas, compared to only one third in the early 1960s. However there are large differences between regions: the rate of urbanisation of the population is only 36 % in ASIA and SSA, whereas it exceeds 75 % in LAM and OECD. In the latter region, which is the second most populated in the world after ASIA (close to one billion inhabitants in 2003), the agricultural working population was divided by almost 3.5 in 42 years and does not exceed 20 million people today.

Food Consumption

Between 1961 and 2003 the world's human population doubled and its food calorie consumption was multiplied by 2.5, reaching approximately 19,000 Gkcal/day at the beginning of the 2000s. Thus, on a global scale, the apparent consumption per person[1] rose from slightly under 2,500 kcal/day in 1961 to just over 3,000 in 2003. However these data mask significant disparities between regions of the world, between the countries of those regions and, within those countries, between and within households. On the scale of MA regions, the mean daily availability was around 4,000 kcal per person in 2003 in OECD, but was still no more than 2,500 kcal in SSA (Fig. 3.3). If we examine these differences we see that they can to a large extent be ascribed to the level of consumption of animal products (milk, meat, eggs, etc.). In OECD nearly 1,200 kcal out of 4,000 (i.e. 30 %) are from animal products, whereas at the other end of the scale, in SSA, these products

[1] Food/Popu: see definitions Chap. 2, this value is often called "food availability".

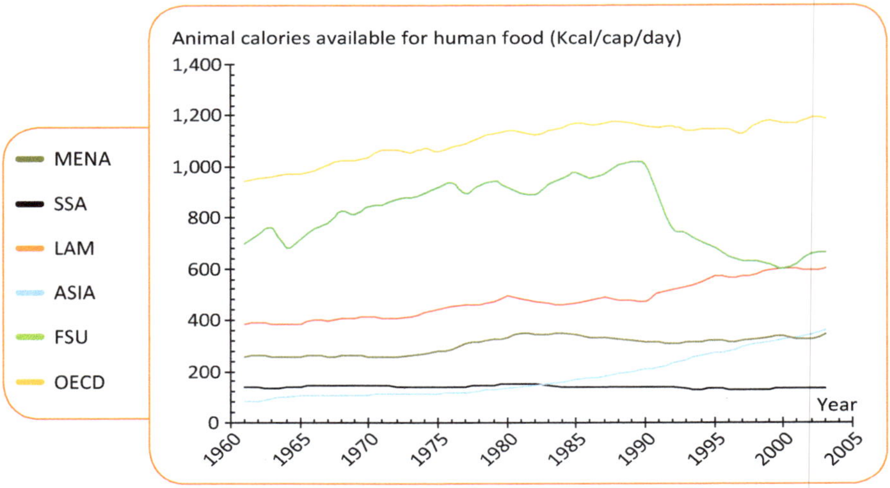

Fig. 3.4 Animal calories available for human food (1961–2003)

contribute to only about 135 kcal out of 2,400 (less than 6 %) (Figs. 3.4 and A2.2). As animal products are particularly rich in protein (meat, eggs, etc.) and/or lipids (butter, cream, etc.), regional differences of food availability are strongly pronounced for these two macronutrients. Our estimations for 2003 are:

- 125 g/person per day of proteins (of which 60 % of animal origin) in OECD, compared to 60 g (of which 20 % animal) in SSA,
- 165 g/person per day of lipids (of which 55 % of animal origin) in OECD compared to 48 g (of which 20 % animal) in SSA.

Land Use

Most of the planet's surface is covered by seas and oceans. Fish and other marine species are concentrated in a small part of these areas, up to a depth of 200 m. These are the continental shelves, covering an area of 2.4 billion ha. Continental shelves are the extended perimeters of the continents, which account for an area of over 13 billion ha. There are no accurate, consistent data on the occupation of the entire area of the continents over the years (Chap. 2), but based on the data and countries considered we know that in 2003 close to 30 % was covered in forests (≃ 3.9 Gha), over 25 % in pastures (≃ 3.4 Gha), nearly 12 % in crops (≃ 1.5 Gha), and just over 3 % in fresh water (≃ 0.4 Gha: lakes, rivers, etc.), with the remaining 30 % (≃ 3.9 Gha) consisting of deserts of varying altitudes and, to a lesser extent, artificialised zones (dwellings, industries, roads, etc.). Between 1961 and

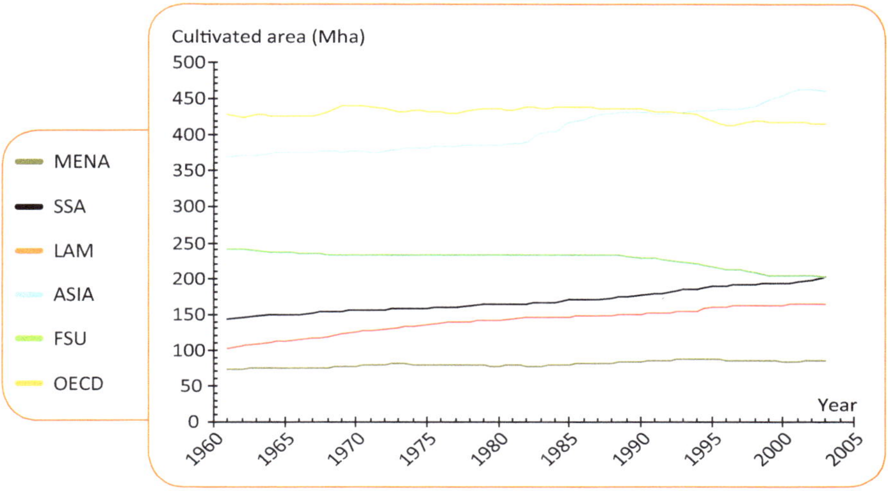

Fig. 3.5 Cultivated area (1961–2003)

2003, cultivated areas had the highest annual growth rate (+0.29% per annum, with approximately +175 Mha over the period), followed by pastures (+0.25%, +330 Mha) and then fresh water (+0.21%, +35 Mha). These increases took place to the detriment of forests (−0.23% per year) with approximately 400 Mha lost in 43 years (Fig. A2.3). In the MA regions, the extension of cultivated areas (Fig. 3.5) was most pronounced in LAM (+1.13% per year), SSA (+0.81%) and ASIA (+0.54%). The first two regions exploit less than 20% of their potentially cultivable land (Fig. A2.4), whereas this rate has been over 80% in ASIA since 1985 (100% in MENA since 1990) (Fig. 3.6).

In contrast, in OECD and especially the former Soviet Union (FSU), cultivated areas have declined (respectively −0.06% and −0.41% annually). With a decrease in the number of farmers, the average cultivated area per farmer is more than 10 ha in FSU and close to 20 ha in OECD (2003) whereas it is less than 4 ha elsewhere, and only 0.5 ha in ASIA (Fig. 3.7).

Food Production and Productivity

Our estimations, based on available data (Chap. 2), show that gross global food production[2] was approximately 33,200 Gkcal/day in 2003, with 62% of this energy from carbohydrates (≃1,883 Mt in 2003), 24% from lipids (≃322 Mt

[2] Plant, animal and aquatic products combined. As some serve in the production of others (e.g. animal feed), this type of addition of products is tricky, just as it is tricky to establish a net balance.

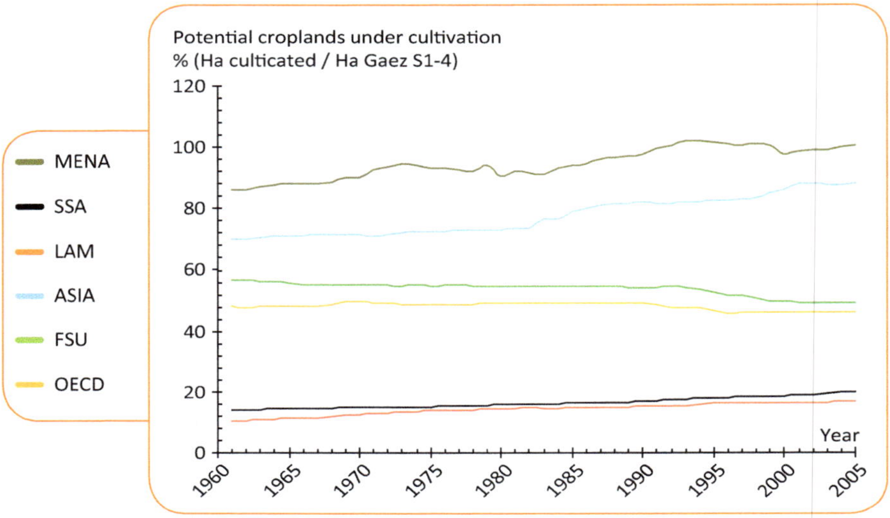

Fig. 3.6 Potential croplands under cultivation (1961–2003). *GAEZ* Global agro-ecological zones

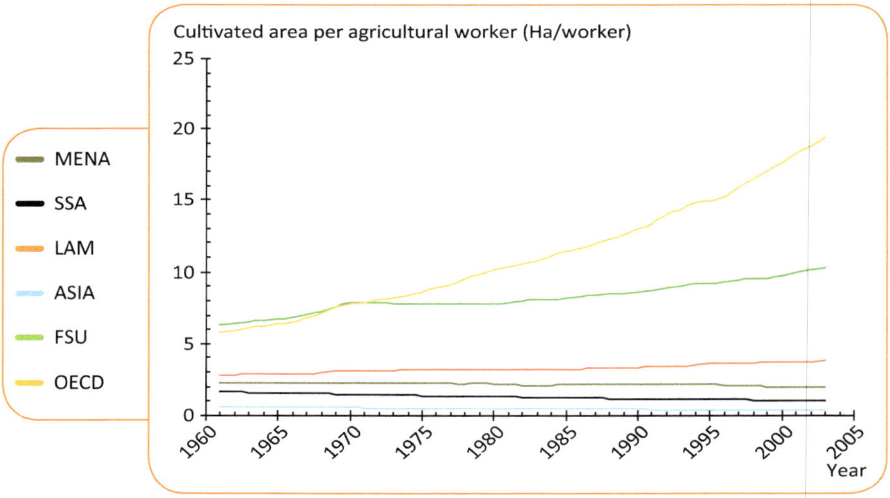

Fig. 3.7 Cultivated area per agricultural worker (1961–2003)

and 14 % from proteins (≃ 420 Mt. Between 1961 and 2003 this calorie production was multiplied by 2.5[3] (Fig. A2.5) without any significant change in the proportion

[3] Between 1961 and 2003, this calorie production (kcal/day) was multiplied by 2.3 for carbohydrates, by 2.7 for proteins and by 3.0 for lipids.

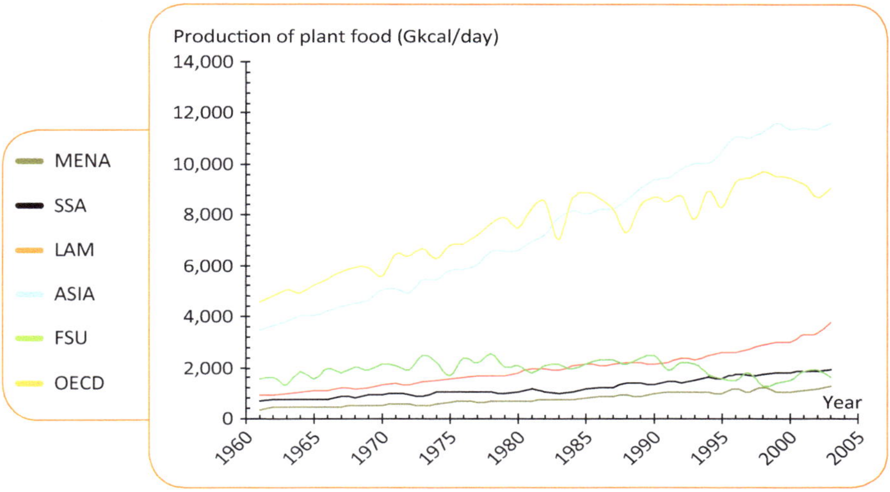

Fig. 3.8 Production of plant food (1961–2003)

of calories from plant, animal or aquatic origin. The order of the producing regions has however changed over the past four decades. During the 1980s, ASIA became the world's leading producer of food of plant origin, thus overtaking OECD (which nevertheless remains the main producer of food of animal origin). LAM took third place, ahead of FSU whose food production is now similar to that of SSA (Fig. 3.8). These production increases are correlated to an extension of cultivated land (see above "land use", p. 58) and, more importantly, to an increase in the food calories produced per hectare cultivated. On a global scale, this production per hectare cultivated (with or without food crops) was multiplied by over 2.2 between 1961 and 2003, increasing from 8,610 to 19,190 kcal/ha a day (from 9 to 21 quintals/year in wheat equivalent). These levels are over 100 times higher than those of aquatic areas which on the whole are not "cultivated" by humans (Fig. A2.6). This growth of food production per hectare of cultivated land can be explained primarily in terms of increasing yields[4] per harvest and per year (increase in the number of harvests per year), owing to greater use of inputs and/or a better combination or use of them (water, fertilisers, seed, pest control, mechanisation, etc.). Today the mean regional production of food calories per cultivated hectare is highest in ASIA (over 25,000 kcal/ha a day). It is also in ASIA that it increased fastest between 1961 and 2003 (+2.35 % per annum), after MENA (+2.7 %) but ahead of LAM (+2.25 %), OECD (+1.7 %), SSA (+1.55 %) and FSU (+0.5 %) (Fig. 3.9). In OECD, the growth of food production per cultivated hectare has gone hand-in-hand with just under a sevenfold increase in the food production per agricultural worker (a regional mean production of almost 425,000 kcal/worker per day in 2003, compared to less than

[4] Rather than by an increase—within cultivated areas—in food crops to the detriment of non-food crops (e.g. fibres, rubber, tobacco or crops for fodder).

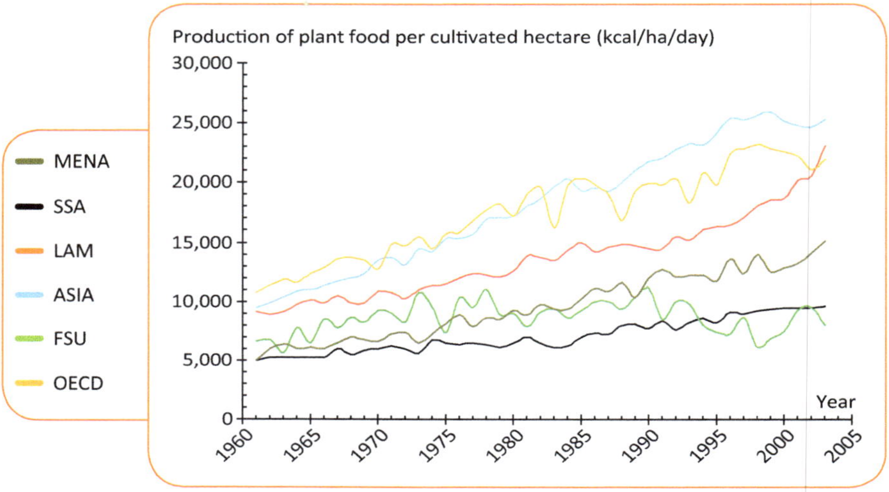

Fig. 3.9 Production of plant food per cultivated hectare (1961–2003)

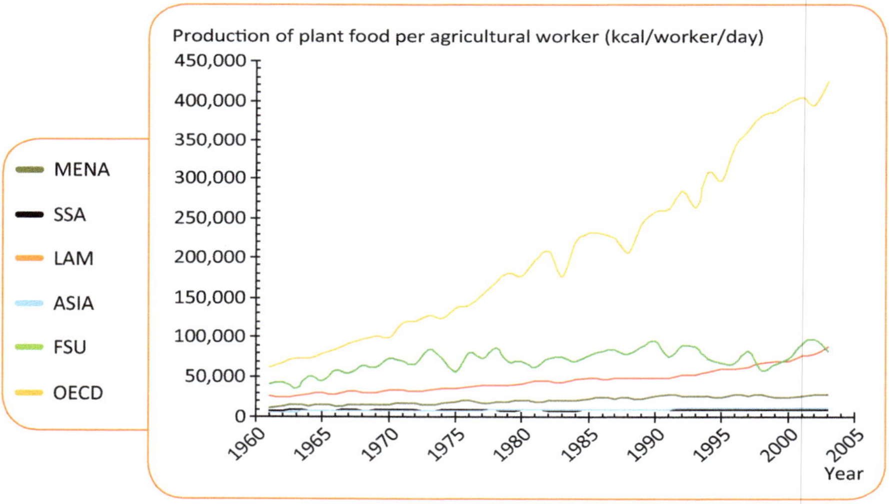

Fig. 3.10 Production of plant food per agricultural worker (1961–2003)

90,000 for the other regions, Fig. 3.10). This rapid growth is a result of the continuation of the rural exodus (Fig. 3.2), the extension of cultivated areas per agricultural worker (Fig. 3.7), and a very high level of motorisation (tractors, harvesters, etc.) as a substitute for human and animal labour.

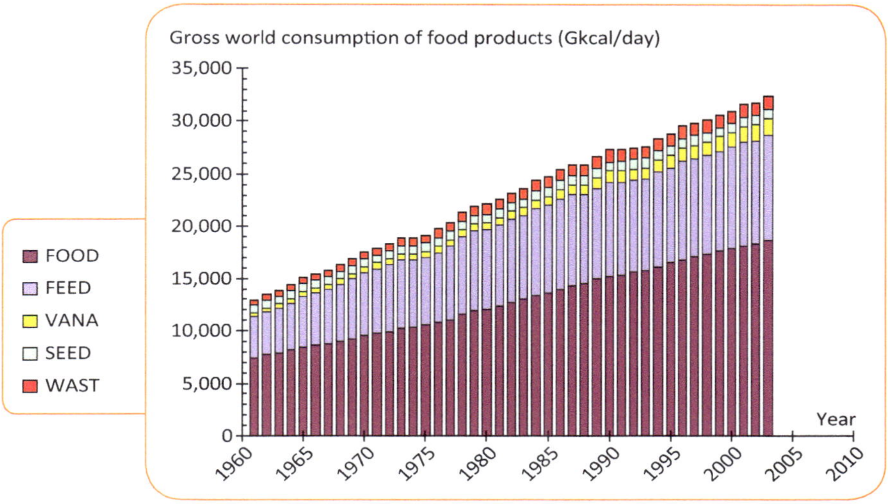

Fig. 3.11 Gross world consumption of food products (1961–2003)

Use of Food Products

The gross world consumption of food products was approximately 33,000 Gkcal/day in 2003, which is equivalent to the total production (see "Food production and productivity", previous page) after taking into account imperfections in the statistics and stock variations (Chap. 2). As for production, over 88 % of this consumption corresponds to plant products, about 11 % to (land) animal products and less than 1 % to aquatic products (essentially animal). These calories are put to various uses, after some waste between production and sale (≃ 4 % in 2003) (Fig. 3.11). Uses include seed and other forms (like eggs for hatching) intended for reproduction (≃ 3 %), as well as non-food agricultural uses (≃ 5 %: lubricants, cosmetics, biofuels, etc.). Most are however used to feed humans and animals, in highly variable proportions, depending on the region, in the case of plant products (Fig. A2.7). In SSA and ASIA, over 70 % of available plant food in 2003 was used to directly feed humans, whereas this rate was only 35 % in OECD, a region that for a long time has devoted over 55 % of its available plant food to feeding animals. This share of plant calories used for feed has been increasing since the early 1960s in LAM, MENA and ASIA where it now ranges between 20 % and 40 % (Fig. A2.8), within specific regional animal production systems that have been modelled (Chap. 2). The share of plant calories used for non-food purposes has also been increasing in most regions, especially since the 1990s and primarily in LAM and OECD where it is now over 5 %.

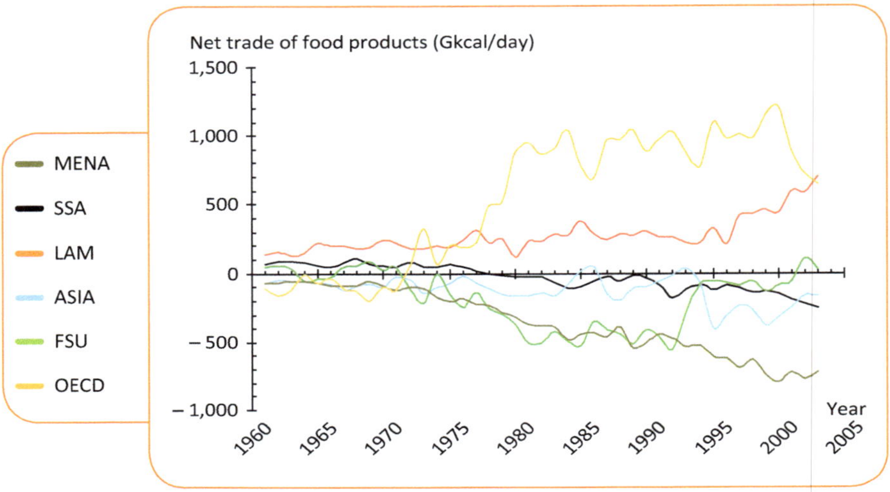

Fig. 3.12 Net trade of food products (1961–2003)

Food Trade

In four decades, the world food trade has grown sharply, from less than 1,500 Gkcal/day in 1961 to over 7,000 Gkcal/day in 2003 (92 % from plants, 6 % from land animals, 2 % from aquatic sources). This growth attests to countries' increasing reliance on international trade, with an attendant growth of transport using fossil fuels. During this period, the direction of international trade in food calories has also evolved considerably. The net trade balances for each of the MA regions (total exports—total imports, Chap. 2) (Fig. 3.12), and their comparison with regional consumption (Figs. 3.13 and A2.9), enables us to draw conclusions on the following five points:

- OECD, a large consumer of food calories, has also become a major exporter of food of plant and animal origin,
- LAM, traditionally a net exporter of food calories (mainly from oilseeds and sugarcane) has remained in the lead and simultaneously increased its average per capita calorie availability,
- ASIA maintains its relative independence in food calories (balance between exports and imports) whereas the regional availability per person has increased, and the number of individuals jumped from 1.5 to close to 3.4 billion between 1961 and 2003,
- export earnings of MENA enable this region to import and consume a growing quantity of food calories, including for the purpose of breeding and feeding animals locally,

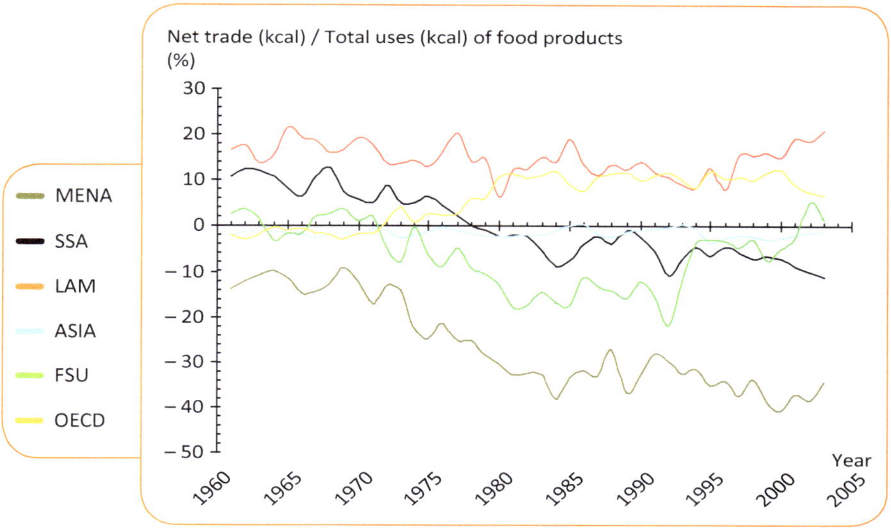

Fig. 3.13 Independence rate in food calories (1961–2003)

- in SSA, local production and net imported quantities have been growing but have not been sufficiently high to substantially increase the calorie availability per person, which has remained the lowest in the world (Fig. 3.3).

Chapter 4
Scenario-Building Choices and Principles

Sandrine Paillard, Tévécia Ronzon and Sébastien Treyer

The proliferation of studies and debates on the future of the global food and agricultural system stems from the increasingly widespread certainty that the continuation of current trends in consumption and agricultural production is encountering limits that only changes in behaviour and technological innovations will enable us to overcome (Collomb 1999; IAASTD 2009; Myers 1991; World Bank 2008).

This certainty has been strengthened by three trends that now appear inevitable: (1) the fact that the world's population is (still) growing fast; (2) climate change; and (3) the increasing scarcity and rising prices of fossil fuels. In view of these trends, several studies have highlighted a possible stagnation of the yields obtained for various crops if current production systems are maintained (IAASTD 2009; Ladha et al. 2003). The Millennium Ecosystem Assessment (MA) (MA 2005c) has furthermore highlighted the deterioration of ecosystems and the consequent threats to the multiple services that they render to humanity.

With these issues at the centre of the analysis, we chose to analyse the MA scenarios, and particularly Global Orchestration, from the point of view of food and agricultural systems. We decided to construct only one new scenario, Agrimonde 1, aimed at exploring the meaning and conditions of existence of a sustainable food and agricultural systems scenario. This chapter explains the reasons for this choice of scenarios, along with the principles underlying the scenario-building process.

S. Paillard (✉)
ANR, 212 rue de Bercy, 75012 Paris, France
e-mail: sandrine.paillard@agencerecherche.fr

T. Ronzon
IPTS, C/Inca Garcilaso, s/n 41092, Seville, Spain
e-mail: tevecia.ronzon@ec.europa.eu

S. Treyer
IDDRI, 27 rue Saint Guillaume, 75337 Paris Cedex 7, France
e-mail: sebastien.treyer@iddri.org

S. Paillard et al. (eds.), *Agrimonde – Scenarios and Challenges for Feeding the World in 2050*, 49
DOI 10.1007/978-94-017-8745-1_4, © Éditions Quæ, 2014

The Agrimonde GO and Agrimonde 1 Scenarios, Built with Reference and by Contrast to the MA Scenarios

The ma scenarios, built to study the future of ecosystems, are references in international debates (Boxes 4.1 and 4.3, and Fig. 4.1). The principles on which they are built are not necessarily the most relevant for discussing the future of food and agricultural systems. It is nevertheless interesting to compare the two approaches, one regarding ecosystems and the other regarding the human activities that have the

Box 4.1–The Millennium Ecosystem Assessment
The Millennium Ecosystem Assessment, a vast operation for international scientific assessment, launched by the United Nations in 2001, took place over a four-year period and mobilised over 2,000 people—mostly experts—from 95 countries. The work, involving several scientific disciplines (biology, ecology, economics, sociology, etc.), was published in several assessment reports intended for policy-makers as well as economic actors and civil society (the MA reports can be consulted on the MA website: www.millenniumassessment.org). The aim of the MA was "to assess the consequences of ecosystem change for human well-being and the scientific basis for action needed to enhance the conservation and sustainable use of these systems and their contribution to human well-being" (Watson and Zakri 2005). Different geographic scales—global, regional, national, local, and watershed—were taken into account and articulated to one another. Nine ecosystems were identified and analysed, from those that are relatively undisturbed, such as natural forests, to ecosystems intensively managed and modified by humans, such as agricultural land and urban areas.

In addition to assessments, the experts built four scenarios of the world up to 2050. The MA scenarios are distinguished by their geopolitical framework (regionalisation versus globalisation) and by the respectively proactive or reactive nature of policies and regulations on ecosystem protection. They are characterised by different societal priorities, especially in terms of poverty alleviation and the protection of ecosystems and natural resources. (Source: MA 2005b)

strongest impact on ecosystems. Moreover, although this objective was not met, the MA scenarios were intended to serve as a basis for building scenarios in the International Assessment of Agricultural Knowledge, Science and Technology for Development (IAASTD) (IAASTD 2009). We therefore chose to analyse the MA scenarios in depth, and to use the Agrimonde platform to apply them to food and agricultural scenarios.

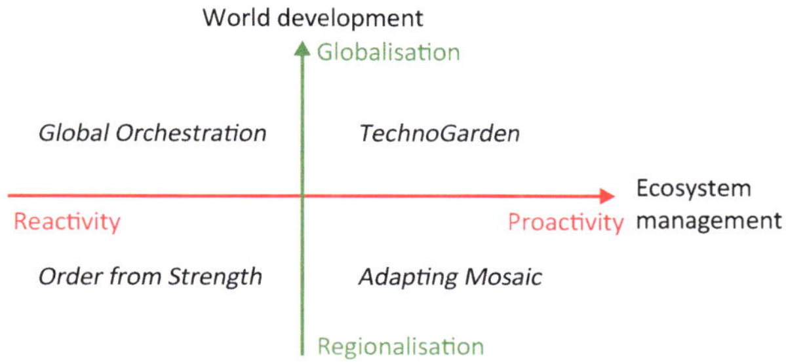

Fig. 4.1 Differentiation of MA scenarios according to the geopolitical framework (regionalisation versus globalisation) and the respectively proactive or reactive nature of policies and regulation on ecosystem management (Source: MA 2005b)

The MA scenarios are exploratory in the sense that they examine the consequences of various possible trends, starting with the present situation. None of them was built as a normative scenario based on a desirable or undesirable image of the future. In *Nourrir la planète* ("Feeding the Planet"), Michel Griffon, following Gordon Conway (Conway 1997), proposed a scenario describing agricultural systems in which all the sustainability characteristics were sought, and the potential and conditions of a "doubly green revolution" were explored (Griffon 2006). This type of agriculture would be characterised by agricultural production technologies liable to both preserve ecosystems and allow for development through agriculture in countries in which the lack of capital limits the implementation of production systems making intensive use of equipment, pesticides and fertilisers.

We were therefore keen to build a desirable scenario, freely inspired by the Doubly Green Revolution (DGR) scenario of Michel Griffon. The idea was to construct a new scenario, starting with the same initial question: what are the technologies of sustainable production in terms of environmental protection and of poverty alleviation through agricultural development? The aim was also to explore possible changes not only in agricultural supply but also in the demand for food and non-food agricultural products.

Box 4.2–The different types of scenario

Scenarios can be distinguished from one another in terms of their positioning in time. An exploratory scenario is built by extrapolating on past and present trends (it starts with the past to imagine the future) while a normative or anticipatory scenario is built retrospectively, that is, starting with a desirable or undesirable vision of the future and tracing back to the present. Scenarios can also be distinguished in terms of the nature of the assumptions on which they are based, relating to the evolution of key variables. A trend-based scenario for instance represents a global assumption of the continuation of current trends and of recently triggered dynamics ("business as usual"), while the contrasting scenario relates to the notion of rupture with the present. (Source: Commissariat général du Plan (2004))

Agrimonde 1 is thus a normative scenario in so far as it simulates a situation in 2050 that differs substantially from today's, and examines the possibility of identifying a path leading there (Box 4.2). It is built with reference to the MA scenarios and that of Michel Griffon, but departs from the former and from today's main trends. Its interest is however not prescriptive but heuristic. Agrimonde 1 assumes that by 2050 the world will be able to implement a sustainable food and agricultural system. The aim is to afford a better understanding of the meaning of such development, with the dilemmas and the main challenges that it entails, as well as the changes and discontinuities that it implies.

In concrete terms, the scenarios of the MA and of Michel Griffon constituted the reference for the development of the Agrimonde 1 assumptions. Additionally, we chose to "reconstruct" a MA scenario, Global Orchestration, so that Agrimonde 1 could be compared to a trend-based scenario on food consumption, but with different underlying societal priorities. Global Orchestration is in fact the MA scenario with the greatest reduction in poverty and malnutrition (Box 4.3). It is based on both the liberalisation of trade and on major technological progress in terms of agricultural yields. The priority given to economic development in this scenario nevertheless results in a mainly reactive management of ecosystems and environmental problems. This scenario was baptised "Agrimonde GO", because it was reconstructed on the basis of the quantification method adopted in Agrimonde (Chap. 2), and also because the population assumptions chosen for this scenario are not those used in the MA. To really be able to compare the Agrimonde 1 scenario with another scenario, it appeared important to make the same "population pressure" assumptions in both scenarios.

Box 4.3–The Millennium Ecosystem Assessment scenarios
Global Orchestration

A worldwide connected society in which global markets are well developed.

The scenario is about global cooperation not only to improve the social and economic well-being of all people but also to protect and enhance global public goods and services (such as public education, health, and infrastructure). There is a focus on the individual rather than the state, inclusion of all impacts of development in markets (internalisation of externalities), and use of regulation only where appropriate. Supra-national institutions are well placed to deal with global environmental problems, such as climate change and fisheries. However, their reactive approach to ecosystem management makes them vulnerable to surprises arising from delayed action or unexpected regional changes. Environmental problems that threaten human well-being (such as pollution, erosion, and climate change) are dealt with only after they become apparent. Problems that have little apparent or direct impact on human well-being are given a low priority in favour of policies that directly improve well-being. People are generally confident that the necessary knowledge and technology to address environmental challenges will emerge or can be developed as needed, just as it has in the past. The scenario highlights the risks from ecological surprises under such an approach. Examples are emerging infectious diseases and other slowly emerging problems that are hard to control once they are established. Other benefits and risks also emerge from the inevitable and increasing connections among people and nations on social, economic, and environmental scales.

TechnoGarden
A globally connected world relying strongly on technology and on highly managed and often engineered ecosystems to deliver needed goods and services.

Technology and market-oriented institutional reform are used to achieve solutions to environmental problems. In many cases, reforms and new policy initiatives benefit from the strong feel for international cooperation that is part of this scenario. As a result, conditions are good for finding solutions for global environmental problems such as climate change. These solutions are designed to benefit both the economy and the environment. Technological improvements that reduce the environmental impact of goods and services are combined with improvements in ecological engineering that optimise the production of ecosystem services. These changes co-develop with the expansion and development of property rights to ecosystem services, such as requiring people to pay for pollution they create or paying people for providing key ecosystem services through actions such as preservation of key watersheds. These rights are generally created and allocated following the identification of ecological problems. Because understanding of ecosystem functions is high, property rights regimes are usually established long before the problem becomes serious. These property rights are assigned to a diversity of individuals, corporations, communal groups, and states that act to optimise the value of their property. We assume that ecological management and engineering can be successful, although it does produce some ecological surprises that affect many people due to an over-reliance on highly engineered systems.

Order from Strength
A regionalised and fragmented world concerned with security and protection, emphasizing primarily regional markets, paying little attention to the common good, and with an individualistic attitude toward ecosystem management.

Nations see looking after their own interests as the best defence against economic insecurity. They reluctantly accept the argument that a militarily and economically strong liberal democratic nation could maintain global order and protect the lifestyles of the richer world while providing some benefits for any poorer countries that elect to become allies. Just as the focus of nations turns to protecting their borders and their people, so too their environmental policies focus on securing natural resources seen as critical for human well-being. But, as in Global Orchestration, people in this scenario see the environment as secondary to their other challenges. They believe in the ability of humans to bring technological innovations to bear as solutions to environmental challenges after these challenges emerge.

Adapting Mosaic
A fragmented world resulting from discredited global institutions.

It sees the rise of local ecosystem management strategies and the strengthening of local institutions. Investments in human and social capital are geared toward improving knowledge about ecosystem functioning and management, resulting in a better understanding of the importance of resilience, fragility, and local flexibility of ecosystems.

There is optimism that we can learn, but humility about preparing for surprises and about our ability to know all there is to know about managing socio-ecological systems. Initially, trade barriers for goods and products are increased, but barriers for information (for those who are motivated to use it) nearly disappear due to improving communication technologies and rapidly decreasing costs of access to information. There is great regional variation in management techniques. Some local areas explore adaptive management, using experimentation, while others manage with command and control or focus on economic measures. Eventually, the focus on local governance leads to failures in managing the global commons. Problems like climate change, marine fisheries, and pollution grow worse, and global environmental surprises become common. Communities slowly realise that they cannot manage their local areas because global problems are encroaching on them, and they begin to develop networks among communities, regions, and even nations to better manage the global commons. The rebuilding is more focused on ecological units, as opposed to the earlier type of management based on political borders that did not necessarily align with ecosystem boundaries. (Source: MA 2005b)

The choice of constructing a scenario with reference to those of the MA and Michel Griffon led to the selection of the same timeline, 2050, and the same geographic zoning into six main regions: Asia (ASIA), the former Soviet Union (FSU), Latin America (LAM), Middle East-North Africa (MENA), OECD-1990 (OECD), and sub-Saharan Africa (SSA) (Fig. 2.2).

Underlying Scenario-Building Principles

The agrimonde 1 scenario was designed to be one of the possible scenarios of sustainable food and agricultural development. The aim was to explore the concept of sustainable development in its classical dimensions (economic, social and environmental). The health dimension of sustainability is also explored through some aspects of its relation to nutrition. As a possible scenario of sustainable food and agricultural development, Agrimonde 1 illustrates a world which by 2050 will have met the challenges of guaranteeing access to healthy food for all and protecting its ecosystems.

The world in 2050 as described in Agrimonde 1 is based above all on sustainable food conditions, allowing for the reduction of inequalities in food and health, through a drastic reduction in both undernourishment and excessive food intake. Moreover, the world in 2050 will have implemented a set of actions to intensify productive systems (i.e. to increase yields per hectare) and to increase production in most regions. These actions will have fulfilled three objectives: satisfying the growing demand; allowing for the growth of income from agriculture in rural areas of the South, and developing environment-friendly agricultural practices. We explored the diversity of possible ecological intensification paths, with particular reference to Michel Griffon's DGR scenario (Griffon 2006) and to his definition of the concept of ecological intensification (Box 4.4). We examined the nature, assets and limits of this type of technological trajectory, according to regional contexts and the chosen objectives.

In order to build a scenario that meets the challenge of guaranteeing access to healthy food for all, while protecting ecosystems, two general principles were applied, not only to the construction of Agrimonde 1 but also to the construction of Agrimonde GO, in a way that facilitates the comparison of the two scenarios:

Box 4.4–The concept of ecological intensification

Ecological intensification consists in increasing yields by using the ecological and biological functionalities of ecosystems to the greatest possible extent. This can be achieved in five main ways:

The first consists in ecological approaches to fertility. These approaches are based on the intensification of the cycle of organic matter, by increasing the proportion of biomass returned to the soil, and by favouring the right humidity and temperature conditions to decompose the biomass. The technique used is that of plant cover crops and plant mulches. In this respect, agriculture is inspired by natural phenomena which ensure the viability and fertility of large permanent ecosystems such as rainforests and pastures. It is also possible to use organic matter, livestock effluent and urban organic waste. Genetic research may furthermore enable the main cereal food crops to fix nitrogen from the air naturally, as legumes do, or to make better use of phosphorus from the soil.

The second way is through water management of an entire ecosystem. The idea is to conserve water to cope with droughts, and to deal with flooding in such a way as to limit erosion and avoid pollution. This implies landscaping according to ecological principles.

The third way is integrated management of major biogeochemical cycles such as the carbon cycle and the nitrogen cycle. Agriculture and forestry are powerful means of carbon sequestration in the soil and in biomass in order to reduce concentrations of greenhouse gases in the atmosphere. Carbon moreover contributes to fertility in the form of organic matter in the soil. Genetic research can also be expected to enable the integration of capacities to contain more carbon or to limit its losses in the main crops. Basically, the combination of plants cultivated in the same area and their rapid succession (agro-forestry, mixed cropping) can facilitate the recycling of mineral elements which would otherwise leave the system.

The fourth way is the integrated control of pests and diseases, especially through organic control. Many pests are controlled by predators, as numerous insects are the prey of other insects or of birds. Ecosystems involve countless highly complex relations between hosts and pathogens (bacteria, microscopic fungi, insects, worms, etc.). Detailed knowledge of these relations makes it possible to use control methods incorporating various strategies, e.g. organic means, targeted chemical means, plant resistance to disease. Genetics can also serve to identify natural ways of controlling populations of pests and systematising them. In the domain of veterinary medicine, "ecopathology" methods are likewise starting to be used, inspired by an ecological approach to disease.

The fifth way is the use of biodiversity. We know that the deterioration of ecosystems (their ecological regression) is attended by losses in biodiversity. The presence in ecosystems of pools of species linked to the environment in complex ways gives them interesting properties: capacities to recycle and to limit losses of nutrients, resistance to disruptions, and high productivity. As knowledge progresses, it will be possible to identify how, by acting on biodiversity, we can contribute to improving productivity and ecosystem resistance to climatic shocks as well as to diseases and pests.

An agriculture that is more productive and more economical in terms of chemical inputs, and that uses ecosystem functioning as a basis for production techniques, thus requires real ecological engineering, a technical domain that integrates knowledge from agronomics—the science on which modern agriculture was based in the twentieth century. (Source: Griffon (2002))

First principle: Assessment of the capacity of each major region of the world to satisfy its food requirements in 2050. This implied that interregional trade would be considered only after an evaluation of the extent to which agricultural production in each of the regions covered local needs. The idea was to focus the analysis on food and agronomic factors and on the actions to implement in order to achieve the above-stated goal. It was therefore necessary, for each of the regions, to explore all possible technological, organisational and institutional trajectories regarding food and agriculture, which would enable the region to guarantee food security by increasing yields and cultivated areas where possible, and by preserving the environment and natural resources. This approach enabled us to identify those areas that would have most difficulty in meeting the food needs of their own population.

Second principle: Analysis of the effects of future demographic trends, without them being masked by large international migratory flows triggered by economic, political or climatic factors. The implications of expected population explosions, mainly in SSA, ASIA and LAM, could thus be fully examined with regard to each region's capacity to feed its own population.

The Agribiom quantitative module, in its present form, nevertheless imposes a number of constraints on quantification and data access. Consequently, certain dimensions of the scenarios have not been quantified or explored in detail, and in some cases certain themes have been excluded.

First, there are no accurate, complete quantitative estimates, in terms of geographical coverage, of the impacts of climate change on the world's agriculture (especially on cultivable areas and yields), and of agriculture on the climate. Consequently, climatic phenomena (greater variability, alterations in rainfall, rising temperatures, thawing of certain areas, etc.) have only been roughly taken into account. The Intergovernmental Panel on Climate Change (IPCC) scenarios show that the inertia of climate change is high and that even the most normative and optimistic scenarios imply an evolution of climatic phenomena by 2050, including a mean global temperature increase of 1 to 2 °C (IPCC 2007a). On the basis of these scenarios, we adjusted our assumptions on cultivated areas and possible yields in 2050 in the different regions. The future impacts of climate change, especially regarding the increasing frequency of extreme events, also called into question the ability of the agricultural systems foreseen in 2050 to withstand such events. The extent of the adjustments that we chose would need to be confirmed by more precise studies on future cultivation potential, in view of climate change. Moreover, to conduct a more robust assessment of the sustainability of each of the scenarios, we lacked data on the respective consequences, in terms of greenhouse gases, of different types of change in land use (pastures or forests converted to croplands, forests converted to pastures, etc.) or of an intensification of agriculture by way of mechanisation and the use of petrochemical fertilisers and pesticides.

Secondly, the quantitative tool Agribiom does not yet integrate indicators of the consumption of natural resources, such as quantities of water or energy consumed[1]. Such endogenisation would have required an in-depth study on the nature of the resources used (e.g. blue water or green water[2]) and their opportunity costs (no allocation possible other than that foreseen in the scenario, another possible allocation, etc.). However, the notion of pressure on natural resources is dominant in the analysis in various respects (deforestation resulting from the extension of farmlands, water stress induced by climatic and demographic changes, deterioration of the quality of the soil and water due to farming practices, etc.). Beyond

[1] The Agribiom module is to be improved in this respect.

[2] The blue water/green water distinction was proposed by Falkenmark in 1995 (Falkenmark 1995). "Blue" water is the water that runs in rivers down to the sea, that is found in lakes, that is captured in the ground, that is distributed in pipes, etc. "Green" water is that which is contained in the soil and is available for plants. 60 % of the total of all precipitation constitute green water.

the identification of the main issues concerning natural resources, analysis turned towards strategies of adaptation and limitation of these phenomena.

Finally, Agrimonde 1 is based on the assumption that agricultural development is a driving force of global economic development and poverty alleviation, as stated in the World Bank Report for 2008 (World Bank 2008). It assumes that in many developing countries the effective food demand and the absence of massive migrations depend above all on the growth of national agricultural production. However, the Agribiom tool is not designed to verify whether the supposed increases in agricultural production in each region do effectively make it possible to contribute to sufficient economic development, to guarantee the supposed increases in food demand and, for instance, to avoid phenomena of massive migrations.

On the basis of these scenario-building principles, quantitative assumptions were formulated on resources and their uses at regional level, for each of the two scenarios. For Agrimonde GO, quantitative assumptions made in the framework of the MA were used, whereas for the Agrimonde 1 scenario rules were designed for the construction of each of the sets of variables to be quantified. Because we wished to grasp the effects of future demographic trends without masking them by major international migratory flows, we chose (for the two scenarios) the United Nations (UN) median projections for 2050, representing a situation of international migrations qualified as 'normal' by the UN, i.e. some 100 million migrants over 50 years (UN 2006). A normative choice, based on an understanding of what a sustainable diet might be, prevailed in the establishment of assumptions on food consumption. Assumptions pertaining to land areas were built on the basis of physical factors of soil availability and quality, compared with sustainability criteria. Finally, the assumptions on yields were formulated by considering them as an adjustment variable of the system. We therefore proposed a range of yields on the basis of past trends and targeted technical progress that would make it possible to preserve the ecosystems. The idea was mainly to test a more optimistic assumption of yields if the low variant did not allow for the generation of sufficient agricultural resources to satisfy needs on a global scale.

Chapter 5
Food Consumption in 2050

Jean-Marc Chaumet, Gérard Ghersi and Jean-Louis Rastoin

The Agrimonde quantitative scenarios are made up of assumptions on regional food biomass resources and their uses, as well as the resource-use balances of each region and, by aggregation, of the world. In order to be able to quantify the uses of agricultural production per region on the 2050 timeline, we developed assumptions at regional level, first for human populations and then for food consumption per inhabitant.

The World's Populations in 2050

The population assumptions of the Millennium Ecosystem Assessment (MA) for 2050 range from eight to nine and a half billion inhabitants, depending on the scenario (Table 5.1; MA 2005b).

Global Orchestration is the scenario with the lowest world population (8 billion in 2050). The demographic transition is relatively fast due to heavy investments in human capital and rapid technological progress, especially in the health sector. Mortality drops to very low levels in the rich countries, in tandem with high economic growth rates and rapid technological progress.

J.-M. Chaumet (✉)
Institut de l'Elevage, 147 Rue de Bercy,
75595 Paris Cedex 12, France
e-mail: Jean-Marc.Chaumet@inst-elevage.asso.fr

G. Ghersi
Maison des Sciences de l'Homme de Montpellier, MSH-Mm Membre du Conseil scientifique, 17, rue de l'Abbé-de-L' Épée, 34090 Montpellier, France
e-mail: gerard.ghersi@msh-m.org

J.-L. Rastoin
SupAgro Montpellier, UMR 1110 Moisa, 2 place Pierre-Viala,
34060 Montpellier Cedex 1, France
e-mail: rastoin@supagro.inra.fr

S. Paillard et al. (eds.), *Agrimonde – Scenarios and Challenges for Feeding the World in 2050*,
DOI 10.1007/978-94-017-8745-1_5, © Éditions Quæ, 2014

Table 5.1 Populations (in millions of inhabitants) in 2000 and in 2050 in the Agrimonde scenarios and the MA scenarios. (Source: see Agribiom, Chap. 2 (based on FAOSTAT, UNSTATS, MA, 2005b))

Region	2000 (Millions of inhabitants)	2050 Agrimonde scenarios (Millions of inhabitants)	2050 (Millions of inhabitants)			
			Global orchestration	Order from strength	Adapting mosaic	TechnoGarden
MENA	353	632	584	691	691	648
SSA	659	1,662	1,128	1,591	1,514	1,349
LAM	515	773	738	939	928	826
ASIA	3,204	4,427	4,005	4,912	4,883	4,430
FSU	281	239	282	257	273	281
OECD	972	1,066	1,136	914	976	1,044
World	5,984	8,800	7,872	9,303	9,265	8,578

In contrast, in Order from Strength, in which the global population attains nine and a half billion in 2050, demographic transition is very slow, due to low investments in human capital and weak economic growth. Unequal growth in rich countries and the emergence of potentially serious diseases results in high mortality rates.

In Adapting Mosaic, the global population in 2050 is practically identical to that of Order from Strength. Both scenarios follow the same trajectory for many years, before an effort is made in education, in Adapting Mosaic, which generates stronger economic growth and boosts technological progress. However, the demographic consequences of these investments, in terms of birth and death rates, take a long time to appear.

In the TechnoGarden scenario the global population is around 9 billion. Moderate investments and economic growth induce moderate fertility and mortality trends in the world, leading to rates qualified as average in comparison with the other scenarios.

The Agrimonde 1 and Agrimonde GO assumptions on the global population in 2050 (Chap. 4) are based on the median projection of the United Nations (UN), which sets the number of the earth's inhabitants at slightly over 9 billion in the mid twenty-first century. This projection assumes a "normal" level of international migration, corresponding to some 100 million migrants over 50 years. This assumption of weak migration—which contradicts certain expectations concerning climate refugees—enables the proposed scenarios to fully reflect the consequences of strong demographic growth anticipated in sub-Saharan Africa (SSA), Asia (ASIA) and Latin America (LAM), without masking them behind major international migratory flows.

The populations per region in the Agrimonde scenarios have been calculated on the basis of figures per country available in the UN projection (Table 5.1). Note that certain countries (Iraq, Afghanistan, etc.) were excluded from the calculations due to a lack of retrospective data (see Chap. 2). We therefore do not take these countries into account in our population figures for 2000 and 2050.

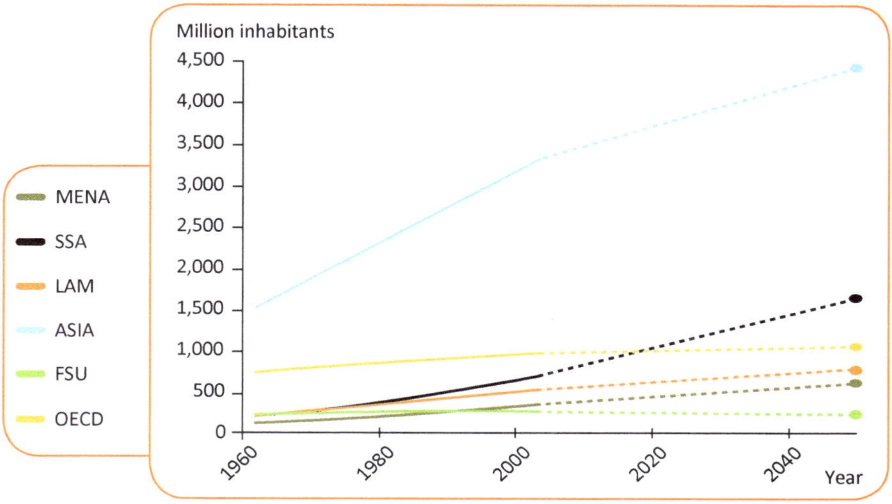

Fig. 5.1 Population trends between 1961 and 2003 and between 2003 and 2050 in the Agrimonde scenarios. (Source: see Agribiom, Chap. 2 (based on FAOSTAT, UNSTATS))

According to the UN median projection, the highest population growth will take place in SSA, where the population will be multiplied by 2.5, followed by Middle East—North Africa (MENA) (multiplied by 1.8), LAM (multiplied by 1.5) and ASIA (multiplied by 1.4). The population of OECD-1990 (OECD) is expected to remain stable, while that of the former Soviet Union (FSU) is expected to drop by close to 15 % (Fig. 5.1).

Diets in 2050

Food Availability in Calories

Global food availability in calorie equivalents per inhabitant rose from 2,500 kcal/cap/day in 1961 to 3,000 in 2003 (Appendix 2, Fig. A2.2). This increase entailed improvements for human societies in the latter half of the twenty-first century: the spectre of famine faded into the background, food prices dropped substantially, food safety was improved, and positive effects were felt throughout the economy (Rastoin 2007). Yet events like the 2008 "food riots" were a reminder that, among other things, about 1 billion individuals in the world are still under-nourished (1.2 billion in 2009, according to the FAO). After declining in the 1980s and early 1990s, this number increased during the 2000s in all regions of the world, with the exception of LAM (FAO 2008, 2009).

In both the Agrimonde and the MA scenarios, "food availability" is used to estimate food consumption (Box 5.1). This availability is measured in terms of the average quantities of calories theoretically available to each consumer. It does not take into account differing consumption levels between households (rich, poor, etc.) and within households (men, women, children, etc.) in each region under consideration. As the calories that are lost between the purchase of products and their ingestion (at home or via collective catering) are included (Appendix 3), food availability should not be confused with the quantities actually ingested, which are far more difficult to estimate regularly and consistently (consumer surveys are costly and protocols tend to change).

Box 5.1—Availability as an approximation of food consumption

In the Agrimonde scenarios, as in the MA scenarios, "food availability" serves as an approximation of food consumption. It is calculated as the ratio between the calorie equivalent of quantities of available foods (production + imports-exports +/-stock variations) to feed the inhabitants of a region (excluding animal feed, non-food uses, seed, and loss after harvesting), over the number of inhabitants of that region. This availability reflects the number of calories available for direct consumption by households or via collective catering. It therefore includes the calories that will be lost between the purchase of products and their ingestion (Appendix 3). It should not be confused with the quantity of calories actually ingested, which is difficult to estimate. In terms of ingestion, the net energy needs of humans are between 2,000 and 3,000 kcal/day, depending on sex, size, weight and intensity of physical activity.

In the MA scenarios on the 2050 timeline, economic growth largely explains the levels of food availability, except in TechnoGarden where the proactive attitude towards the environment limits the increase in animal product availability (Box 5.2). Hence, we may consider that the Agrimonde GO scenario, which applies the assumptions of the MA Global Orchestration scenario (Appendix 3), is a trend-based scenario in terms of the evolution of the consumption of food calories. Economic growth drives consumption in all the regions of the world up to a mean global availability of 3,590 kcal/cap/day: from close to 3,000 kcal/cap/day in SSA to around 4,100 kcal/cap/day in OECD (Fig. 5.2 and Table 5.2). Significant differences are still found in the consumption of plant, animal and aquatic products, despite the global increase in wealth (Fig. 5.4). Under-nourishment is reduced considerably but modes of consumption in countries with fast economic growth cause a parallel increase in diseases due to over-nourishment (cardiovascular diseases, overweight or obesity, etc.).

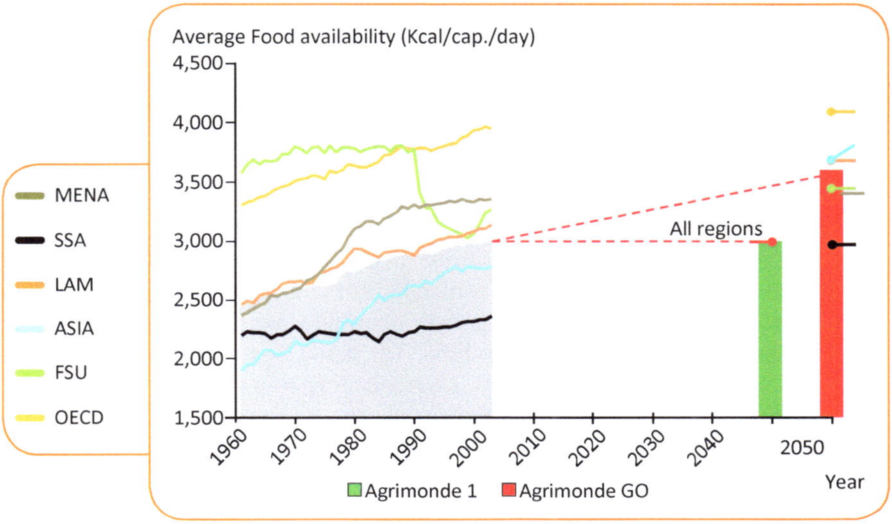

Fig. 5.2 Average regional food availability from 1961 to 2003 and in 2050 in the Agrimonde scenarios

Box 5.2—Assumptions of food consumption in the MA scenarios

Global Orchestration

The mean global food availability in 2050 is 3,600 kcal/cap/day. This scenario is a continuation of current trends, with an extension of the obesity epidemic in areas with strong economic development and a steep increase in the demand for meat. The corollary is a growing proportion of cereals used for animal feed. Traditional models are gradually abandoned as food habits become more standardised. The number of calories available for food is higher than in the other three scenarios. This continuation of trends allows for significant progress to be made in terms of food security, and for a reduction in absolute terms of the number of children suffering from malnutrition.

TechnoGarden

This scenario differs considerably from the others in that consumers have a proactive attitude towards environmental management. As a result, there is an increasing demand for products from "eco-friendly" agriculture and a moderate increase in the consumption of meat. The increase in the quantity of animal products consumed by the inhabitants of developing countries is similar to that of scenarios centred on regionalisation. On the other hand, in the rich countries the increase in the consumption of animal products is two to three times lower than in the other scenarios. This evolution of diets is linked to the fact that their environmental and health effects are taken into account. Daily global consumption per capita is close to 3,300 kcal in 2050. The number of children suffering from malnutrition is relatively low in this scenario, but higher than that foreseen in Global Orchestration.

Order from Strength
The Western dietary pattern neither evolves in developed countries nor spreads in developing countries. Trade barriers hinder its generalisation, along with weak economic growth which limits the increase in income that it would require. Limited economic growth likewise affects the rich countries, where consumption increases very little. The result is a mean global availability close to 3,000 kcal in 2050, corresponding to a stationary situation throughout the world. Relative economic stagnation, combined with the lack of international cooperation, leads to an increase in the number of children suffering from malnutrition.

Adapting Mosaic
The problem during the first years of this scenario is similar to that of Order from Strength, where economic growth is also weak. Products with a high value added are not affordable for many consumers, who consequently focus on goods produced locally by means of production techniques suited to the local environment. From 2020, the economic situation improves and the demand for healthy, good quality foods increases considerably, to the extent that several countries enshrine "the right to healthy food" in their constitution. This tendency first appears in developed countries, before spreading to the developing world. The mean global food availability in 2050 is slightly over 3,000 kcal/cap/day. (Source: MA 2005b)

The Agrimonde 1 scenario shows a distinctly different picture. In some respects it correlates to TechnoGarden, as in both scenarios health- and environment-related concerns cause the relationship between income and food consumption to change. In Agrimonde 1, the assumption is that food availability is equivalent to 3,000 kcal/cap/day in 2050 in all the regions of the world, i.e. the global average in 2000. An estimated 500 kcal are of animal and aquatic origin, although the composition (in products from grazing animals and non-grazing animals, and from fresh water and marine sources) varies according to certain regional particularities. This convergence of all the regions towards average food availability of 3,000 kcal/cap/day in 2050 represents a major shift compared to the trends observed from 1961 until the early twenty-first century. It involves only small changes in food availability per person in most of the regions until 2050, except in SSA, where we assume that per capita food availability has increased by 30 % in 50 years, and in the OECD region where it has declined by nearly 25 % (Fig. 5.2).

Table 5.2 Total food availability and distribution according to sources in kcal/cap/day in 2000 and in 2050 in the Agrimonde scenarios

Region		Food availibility in 2000 (kcal/cap/day)	Food availibility in 2050, Agrimonde scenarios (kcal/cap/day)	
			Agrimonde 1	Agrimonde GO
MENA	Total food availibility	3,339	3,000	3,457
	Plant	2,990	2,500	2,987
	Animal	334	460	442
	Aquatic	15	40	28
SSA	Total food availibility	2,323	3,000	2,972
	Plant	2,176	2,500	2,667
	Animal	133	479	283
	Aquatic	14	21	22
LAM	Total food availibility	3,090	3,000	3,698
	Plant	2,470	2,500	2,758
	Animal	601	458	892
	Aquatic	19	42	48
FSU	Total food availibility	3,062	3,000	3,457
	Plant	2,439	2,500	2,091
	Animal	599	463	1,296
	Aquatic	24	37	70
ASIA	Total food availibility	2,812	3,000	3,703
	Plant	2,426	2,500	2,766
	Animal	326	412	871
	Aquatic	30	88	66
OECD	Total food availibility	3,941	3,000	4,099
	Plant	2,782	2,500	2,385
	Animal	1,167	451	1,628
	Aquatic	46	49	86
World	Total food availibility	2,991	3,000	3,588
	Plant	2,485	2,500	2,698
	Animal	477	438	834
	Aquatic	29	62	56

Composition of Total Food Availability According to the Source

Diets today still vary widely between regions and countries of the world, in terms of total calorie intake—as seen above—and distribution between the different food sources. The main difference between the consumption models of countries of the South and those of the North is the consumption of animal products: meat, eggs and milk. Western countries have an average availability of three times more meat per capita per annum than countries of the South with a market economy (a North American has an availability of eight times more meat than an African, twice as many eggs and six times more milk). The Western dietary pattern is also far richer in animal products than that of FSU and Eastern Europe.

In the quantitative tool Agribiom, total availability consists of availability of calories from plants, land animals (non-grazing and grazing animals) and aquatic products (fresh water and marine). While the average consumption of plant calories increased between 2000 and 2050 in the Agrimonde GO scenario (+9% on average), the increase in food consumption in this scenario is accounted for essentially by the consumption of calories from animals and aquatic sources (+76% on average) (Fig. 5.3).

The Agrimonde 1 scenario assumes that the availability of plant calories converges towards 2,500 kcal/cap/day on the 2050 timeline in all the regions of the world. This figure corresponds roughly to the global average of plant kilocalories per capita in 2000. Plant food availability in all the regions of the world, with the exception of MENA, tends towards the same figure in 2000 (Fig. 5.3).

The distribution of the 500 kcal/cap/day of animal and aquatic origin (between grazing animals, monogastric animals, fresh water and marine organisms) in the Agrimonde 1 scenario varies from one region to another, since it is through this distribution that we chose to take into account the—mainly cultural—specificities of diets. To determine this distribution, we first established the number of calories from grazing animals (milk and meat) by applying the proportion of calories from grazing animals recorded in 2003 (the last year of observation available in Agribiom), in each region, to 3,000 kcal. The quantity of kilocalories obtained was limited to 250 kcal/cap/day so that it did not exceed half of the total animal and aquatic calories available for food in 2050, (i.e. 500 kcal/cap/day). Thus:

- for regions where the volume was limited to 250 kcal/cap/day in 2050, the availability of calories from grazing animals dropped by about 60% compared to the year 2000 in OECD, by 40% in FSU, and by close to a third in LAM,
- in the other regions, between 2000 and 2050, the availability of calories from grazing animals increased by a quarter in SSA, to attain 129 kcal/cap/day, and by close to a fifth in ASIA, to attain 159 kcal/cap/day, whereas it dropped by nearly 10% in MENA, to 222 kcal/cap/day.

It was then assumed that there was an increase (compared to 2003) in the share of calories of aquatic origin, based on a global historical trend. Differences in the progression for each region were introduced, stemming from regional production capacities as reported by the expert panel. Generally speaking, it was considered that the oceans represent a considerable source of food, but that fishing would be confronted with structural limits related to several factors (over-fishing, artificialisation of the shoreline, pollution, accelerated erosion of biodiversity), while tensions over the uses of fresh water would worsen.

It was assumed that marine aquaculture could increase faster than over the past 40 years, but at different paces:

- high in ASIA, OECD and LAM, with a twofold increase in the growth rate of availability observed over the period 1961–2003,

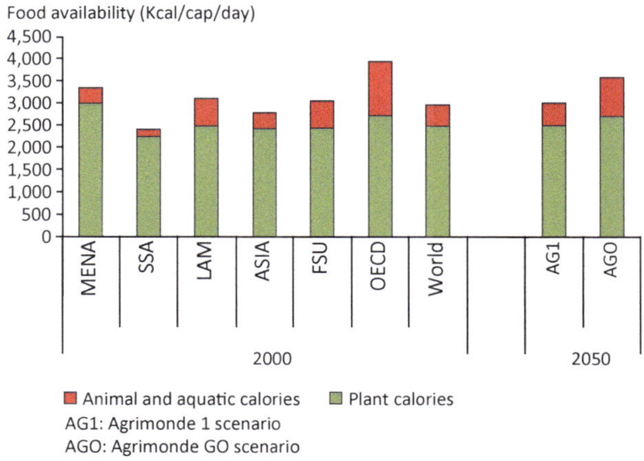

Fig. 5.3 Regional distribution of food availability in products of plant, animal and aquatic origin in 2000 and 2050 in the Agrimonde scenarios

- moderate in the other regions (FSU, SSA, MENA[1]) with an increase between 2000 and 2050 of only 50 % of the growth rate observed during the period 1961–2003.

Given the tensions over freshwater resources, the per capita availability of calories from freshwater fish was aligned with population growth in each region and on this basis was assumed to be relatively stable.

Finally, the availability of calories from monogastric animals was deduced by calculating the difference between total regional availability of animal and aquatic products (set at 500 kcal/cap/day) and the sum of regional availability of calories from grazing animals and aquatic products.

Thus, in the Agrimonde 1 scenario, the growth in the availability of animal and aquatic calories in certain regions (MENA, SSA and ASIA), between 2000 and 2050, stems essentially from the increase in availability of products of monogastric origin. The proportion of grazing animals nevertheless remains high in the remaining regions: OECD, FSU, and LAM (Fig. 5.4).

Agrimonde 1 Scenario: A Set of Assumptions Exploring the Possibility of Major Changes in Diet

In the Agrimonde 1 scenario, the assumptions on diet are markedly different from past trends. This is a set of very strong assumptions since they imply that multiple actors (consumers, producers, processors, retailers, public policy-makers, etc.) take

[1] The calculations for FSU were based on the period 1993–2003, as availability declined in the period 1963–2003. This decline mainly occurred during the period 1989–1993, after which an upward trend was recorded.

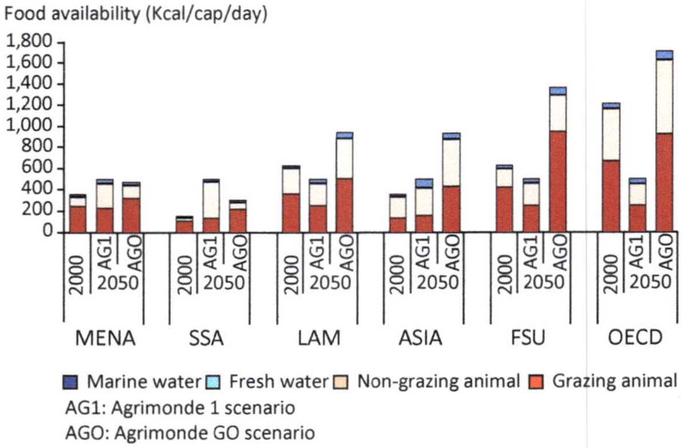

Fig. 5.4 Distribution of calories of animal and aquatic origin in 2000 and in 2050, in the Agrimonde scenarios

into account the global and local environmental and health-related consequences of modes of food production and consumption.

The choice of this set of assumptions is based on four main arguments:

- today's very marked discrepancy between observed availability and the availability required for food security in each region,
- the importance of equity in a sustainable development scenario,
- the relationship between health and food,
- the relationship between diet and pressure on natural resources.

Noteworthy Discrepancy Between Actual Availability and the Availability Required for Food Security

According to the FAO, depending on the inequality of access to food and the heterogeneity of food rations within the population, and assuming that consumer waste is limited (waste that takes place after food reaches consumers, Appendix 3, p. 261), an average availability of 3,000 kcal/cap/day would make it possible on the scale of a population to maintain the proportion of under-nourished individuals at a relatively low level (of approximately 6% of the global population if inequalities are substantial) (FAO 2003a, b).

The assumption of 3,000 kcal/cap/day of average food availability in all the regions in 2050 thus has the advantage of highlighting existing discrepancies at the beginning of the twenty-first century, between the availability needed for food security and actual observed availability, especially in countries with a generalised "Western" consumption pattern. Average availability per capita and per day in 2000 is close to 4,000 kcal in the OECD area and attains almost 4,500 kcal in the United States.

These differences can be explained by the variety of diets within the populations considered, by over-consumption in rich countries, and by a large proportion of

Table 5.3 Distribution of the world's population and of the consumption of plant food calories (2003)

Region	Population	Plant calorie consumption
World	5,973 billion	29,341 Gkcal/day
MENA	6%	7%
SSA	11%	7%
LAM	9%	11%
ASIA	54%	40%
FSU	4%	5%
OECD	16%	30%

waste after food has reached consumers. This waste at the consumer level is worst in developed countries. The figures vary from one survey to another but are mostly high. A study undertaken in the US, for instance, estimates that 14% of the food (meat, cereals, fruit and vegetables) bought by consumers is wasted (Jones 2004), while a study in the UK estimates waste among consumers at a third of all food purchased, 61% of which could be avoided (WRAP 2008; Appendix 3, p. 261).

The Question of Equity in a Sustainable Development Scenario

In the Agrimonde 1 scenario, the issue of greater equity of food consumption, especially between countries of the North and those of the South, was also important in the establishment of assumptions on diets. Consumption today is very unequal, particularly since certain regions use far more plant calories than others for animal feed (Fig. A2.7, and Table 5.3).

The question of equity is also important in Philippe Collomb's book *Une voie étroite pour la sécurité alimentaire d'ici à 2050* (Collomb 1999), however it gives rise to less demanding assumptions than in the Agrimonde 1 scenario. The author assumes that in 2050 the populations of developing countries will enjoy a diet similar to that of Mexico in the late 1990s (3,040 kcal/cap/day of available food). In other words, the countries currently situated below this level will see their food availability increase, whereas those situated above it, especially developed countries, will witness no change in this respect.

Taking the Relationship Between Food and Health into Account

An average availability of 3,000 kcal/cap/day can have positive effects in terms of public health. First, as noted above, the FAO argues that this level of availability maintains the probability of under-nourishment at a relatively low level. Secondly, over-consumption is a cause of nutrition-related non-communicable diseases such as obesity, whose increasing incidence triggers change in food consumption patterns. Studies show that the health-related effects of current diets affect life-expectancy in countries such as the United States (Olshansky et al. 2005). Obesity is the main disease concerned, but not the only one; others include Type-2 diabetes, cardio- or cerebral-vascular diseases, and certain cancers. Obesity is often a risk factor of these diseases but they are more frequently linked to poor diet. Even though Agrimonde 1 does not put forward quantitative assumptions on the nutrients present in diets (Agribiom simulations are performed on calories and not on macro-nutrients and micro-nutrients), it assumes that diets will be more balanced than they are today, from a nutritional point of view.

Limiting the Pressure on Natural Resources

Excessive calorie consumption generates health-related problems as well as negative environmental externalities. The objective of adequately feeding 9 billion inhabitants in 2050 implies *a priori* greater pressure on natural resources than there already is today (e.g. need for land, water and energy), irrespective of the volume and methods of production envisaged. In Agrimonde 1, assumptions on diet also aim to explore and discuss the relationships between diet and pressure on natural resources.

In Agrimonde 1, the availability of food from fresh water sources, for example, increases very little because the scenario takes into account tension and pressure which already exist on these resources in many regions of the world. The assumption of reduced availability of animal products, especially from grazing animals in OECD, and a limited increase in the other regions, also stems from the environmental and energy impact of livestock breeding. For environmental reasons, the MA TechnoGarden scenario also envisages changes in meat consumption in several regions of the world.

From a nutritional point of view, animal calories account for only 16% of all available calories today at global level, but from an economic and environmental point of view their importance is greater. In other words, plant calories (PC) serve to produce animal calories (AC) or directly provide final calories (FC) consumed by humans. When we measure the calorie level of an individual's or population's food ration, we can take into account all the plant calories necessary for the production of final calories. In this case we talk of initial calories (IC), and the food ration (expressed in IC) can be calculated as follows: $IC = PC + (AC \times TC)$, where TC is the transformation coefficient giving the number of plant calories necessary to produce an animal calorie. In the literature these coefficients generally range from 4 to 14, depending on the animals concerned.[2]

In Agrimonde, via Agribiom, this transformation of plant calories into animal calories has been used specifically to model this process in finer detail than in the

[2] In the case of pork, it takes 4 kg of feed in the form of grain to produce 1 kg of meat for consumption. The rate of feed conversion is inferior for chicken and far greater for beef. This means that animals consume approximately 400 kg of grain of all kinds, which they then transform into 80 kg of meat, representing our annual consumption (Malassis and Padilla 1986).

above formula (Chap. 2). These animal production models are designed to estimate the plant products required for food production, as well as needs in terms of land (cultivated and pastures), water, inputs and other production factors.

In general, the study of the consumption of plant calories for food shows that globally almost a quarter is used to feed animals (Fig. A2.7), with a coefficient of conversion of these plant calories into animal calories greater than three, on average. This coefficient varies widely according to time and place (Fig. 2.4), depending on the other resources available for animal feed (fodder, pastures, harvest residues, food residues, etc.).

Even though the production of animal calories requires large quantities of plant calories, caution is nevertheless called for when it comes to environmental impact. It may be considered advantageous to produce animals which optimise the use of plant resources (they consume fibres that humans cannot digest). However, production systems have evolved over the past four decades and the response to the increased demand for beef has been intensification. This has generally led to the reduction of pastures and increasing use of concentrates, especially feed. Hence, while the production of grazing animals in the world increased by 40% between 1970 and 1995, pasture areas grew by only 4% (Bouwman et al. 2005). However, according to these authors, grazing animals' efficiency in converting starch and oilseed crops into meat is lower than that of monogastric animals. A study has shown that in certain beef-production systems, the cultivated area required is three times larger than that used for the same production of monograstric animals (Wirsenius 2003). This conclusion cannot however be generalised to all production systems, least of all those of developing countries.

Advantages in the production of grazing animals still lie in the fact that value can be extracted from land that is often unsuitable for crops (high altitudes, steep slopes, semi-arid areas, etc.), and in the carbon storage and biodiversity conservation by such land. However their production also generates greenhouse gases such as methane, carbon dioxide and nitrogen protoxide (Steinfeld et al. 2006), either directly (breathing, grazing) or indirectly (animal feed, processing, transport). Indirect impacts are increasing with the intensification of production. Finally, grazing animals have various uses: they are a form of wealth for their owner, they produce organic fertiliser and are often used for traction, they provide food in the form of milk or meat, and they are a source of regular income for populations among the poorest in the world in economic terms.

The amount of water required for food production also varies widely, depending on food type: 100 L for 1 kg of potatoes; 4,100 kg for 1 kg of chicken meat; 4,600 L for 1 kg of pork; and 13,000 L for 1 kg of beef (Zimmer and Renault 2003). These figures are of course indications (results obtained in California) and will vary depending on the soil, climate and production system, especially when pigs are fed residues. They do however give an indication of the difference in the amount of water needed for the production of various types of food.

From the point of view of energy, the conversion from the production of animals that graze on local pastures, to the production of monogastric animals requiring the production and transport of feed over long distances, generates considerable additional energy expenses (Steinfeld et al. 2006). In the United States, for example, 2,700 kg of fossil energy is required to produce 100 kcal of pork, and 1,600 kcal to produce 100 kcal of beef (Pimentel and Pimentel 1996).

On a more general note, the production, manufacture and marketing of food consume high levels of energy. The entire food chain reportedly accounts for 17% of the total fossil energy consumption in the United States (Horrigan et al. 2002). The system is furthermore reported to use 100 kcal to produce 7.6 kcal actually purchased by consumers (Eshel and Martin 2006). Food over-consumption thus leads to an increase in fossil energy use which is detrimental to the environment.

Chapter 6
Land Use in 2050

Tévécia Ronzon

In order to quantify food resources in the Agrimonde scenarios for each region of the world, quantitative assumptions had to be made on land use, enabling us to define the areas of food production, the food crop yields, and the conversion of plant calories into animal calories. This chapter is devoted to these quantitative assumptions on land occupation. First, the quantification principles adopted will be presented, followed by the assumptions that were formulated for each region and therefore for the world.

We refer to the data presented in the overview of the world food economy from 1961–2003 (Chap. 3).

The Principles for Quantifying Land Use

There are two predominant approaches in scenario analysis that quantify cultivated surfaces.

In the first approach, evolutions of arable land are obtained by modelling an equilibrium of food supply and demand. The quantification of cultivated areas thus depends both on the level of food demand and on that of the other parameters influencing the offer, such as yields, surface of irrigated land, irrigation efficiency[1], cropping intensity, the level of deterioration of soil fertility, the conversion of farmland into urban areas, etc. This approach is used in studies quantifying scenarios on the basis of the IMPACT model, such as the Millennium Ecosystem Assessment

[1] "The efficiency of irrigation within a perimeter is the relationship between the volume of water drawn or pumped at the head-end, and the quantity actually used by crops (evapo-transpiration), to which one must add the needs related to maintaining a sheet of water in the case of rice" (CIRAD-GRET-MAE 2002).

T. Ronzon (✉)
IPTS, C/Inca Garcilaso, s/n 41092, Seville, Spain
e-mail: tevecia.ronzon@ec.europa.eu

S. Paillard et al. (eds.), *Agrimonde – Scenarios and Challenges for Feeding the World in 2050*, 73
DOI 10.1007/978-94-017-8745-1_6, © Éditions Quæ, 2014

(MA) (MA 2005b), various studies by the International Food Policy Research Institute (IFPRI) (Von Braun et al. 2005; Rosegrant et al. 2002, 2001) and a number of other studies on more specific topics such as the evolution of biofuel crops (Hoogwijk et al. 2005). It is also developed in other studies, using different models, such as in the European Environment Agency's PRELUDE scenarios for Europe (EEA 2007). A second approach consists in making assumptions on the evolution of arable land, based on an estimation of the potential for growing crops. This is the case of projections of the Food and Agriculture Organization of the United Nations (FAO) on agriculture (FAO 2000, 2003a) and the Doubly Green Revolution (DGR) scenario proposed by Michel Griffon (Griffon 2006).

In the MA scenarios, the extension of cultivated surface areas is one of the components of the equation aimed at balancing food production and demand. Assumptions on future cultivated surfaces vary according to the level of other variables: low in Global Orchestration since the population size is low and the degree of agricultural intensification is high, moderate in Adapting Mosaic and TechnoGarden as the demand for meat products is low, and high in Order from Strength because the demographic pressure is high and yield levels are low. Thus, the higher the yields and the smaller the population, the smaller the cultivated surface areas will tend to be. All MA scenarios therefore foresee an increase in cultivated areas, but in different proportions and for different reasons (Box 6.1).

Box 6.1—Assumptions on cultivated areas in the Millennium Ecosystem Assessment scenarios

Global Orchestration

In 2050 most of the newly-cultivated areas are devoted to energy (biofuels). The small proportion of newly-cultivated land allocated to food crops (5 %) can be explained by the sharp increase in yields, primarily as a result of rapid technological progress and investments in agricultural research. LAM and SSA are the regions with the largest areas of newly-cultivated land in 2050. On a global scale, irrigated areas increased by 0.18 % annually between 2000 and 2050, from 239 to 262 million ha. This increase has required considerable investments in irrigation systems. LAM is the region with the fastest growth of irrigated areas (+0.5 % per annum), followed by SSA (+3.0 per annum). The moderate increase in cultivated areas has allowed for the development of biofuels, whose production has been stimulated by rising energy prices due to rapid economic growth. Most of the newly-cultivated land devoted to biofuels is in the OECD region and LAM.

In the Agrimonde GO scenario we adhered, as far as possible, to the evolution of land use foreseen in the MA Global Orchestration scenario. The MA experts had not however based their assumptions of land use on the same statistical data

TechnoGarden

This scenario differs from the others in that the total agricultural areas have increased between 2000 and 2050, but the cereal crop areas have shrunk. The main reasons are: yield increases that have made the expansion of cultivated areas towards marginal lands unnecessary; improved land use owing to new technologies applied to existing land; and an increase in biodiversity conservation programmes. Compared to the beginning of the century, cereal crop areas in 2050 have decreased by 10% in OECD, 7% in FSU, and 6% in ASIA. On the other hand, they have increased by 37% in SSA, 9% in LAM and 7% in MENA. Irrigated areas total 253 million ha in 2050, which is the highest figure after the Global Orchestration scenario. The growth of these areas is steep in LAM, SSA, and MENA.

Order from Strength

Between 2000 and 2050, low yield increases, high demographic growth rates, and the maintenance of trade barriers such as customs duties and quotas have necessitated the use of larger areas for food production. Irrigated areas have remained stable: at global level they have increased from 239 to 240 million ha. Notwithstanding this overall stability, the situation has evolved differently in each region: while irrigated areas have decreased in ASIA and FSU, they have increased in LAM and SSA. In many countries economic difficulties have led to a substantial reduction in investments in irrigation systems. Intense competition has developed between energy production and food production, which has pushed up the cost of biofuels. Meanwhile, slow economic growth has moderated energy demand. These two factors explain the weak growth of areas reserved for the production of biofuels (the weakest of the four scenarios).

Adapting Mosaic

Even though the growth in demand for food from 2000 to 2050 has not been as high as in other scenarios, the slow growth of yields has led to a significant increase in agricultural areas. In 2050 SSA, LAM and MENA have the majority of newly-cultivated cereal crop areas. Irrigated areas have on the whole grown very slowly (0.06% per annum), to a total of 246 million ha in 2050, although they have increased considerably in SSA and LAM.

Adapting Mosaic is an intermediate scenario: economic growth and agricultural productivity are greater than those foreseen in Order from Strength, but less than those in the other scenarios. Consequently, the energy demand has proved to be greater than that measured in Order from Strength, but competition for land has been less intense. In 2050 the areas devoted to energy and biofuels in particular are larger than those of Order from Strength but smaller than those of TechnoGarden. (Source: MA 2005b)

as those used in Agribiom (which were FAO data). The quantitative assumptions made in the MA Global Orchestration scenario were therefore adapted to make them comparable with the assumptions in Agrimonde 1. As a result, in some cases the evolution of land use by region in Agrimonde GO differs slightly from that of Global Orchestration (Appendix 4, p. 263 and Table 6.1 for further details).

In the Agrimonde 1 scenario, the formulation of quantitative assumptions on cultivated areas in 2050 is based on a different logic to that on which the MA scenarios were based. Physical factors of land availability and quality guided the formulation of Agrimonde 1 assumptions. They were analysed in relation to sustainability criteria, such as the preservation of forests, which provide numerous ecological services.

We first sought to identify the reserves of cultivable land in each region. The various types of land areas in the Agrimonde 1 scenario in 2050 were then quantified region by region—cultivated areas, pastures and forests—and the distribution of land between them. The parts of the cultivated areas that would be irrigated, and/or devoted to biofuel crops were finally specified. The development of the regional scenarios on land use proceeded in three stages:

- the identification of potentially cultivable land reserves,
- the quantification of newly-cultivated land and new areas irrigated in 2050, and consequently the remaining forests and pastures,
- the specific quantification of areas devoted to the production of biofuels.

The identification of potentially cultivable land reserves in each region, which could allow for a future increase in regional production, was based on estimates of cultivable land in each region, by the FAO (2003a) and by Fischer et al. (2001, 2000, 2002) (Chapter 2, p. 33). Through their GAEZ (Global Agro-Ecological Zones) methodology, the latter propose a typology in which five levels of cultivation potential are distinguished, in relation to the soil's yield potential:

1. Areas of land whose yield potential is greater than 80% of the maximum attainable yield: VS (Very Suitable),
2. Areas of land whose yield potential is greater than 60% of the maximum attainable yield: VS + S (S: Suitable),
3. Areas of land whose yield potential is greater than 40%: VS + S + MS (MS: Moderately Suitable),
4. Areas of land whose yield potential is greater than 20%: VS + S + MS + mS (mS: marginally Suitable),
5. Areas of land whose yield potential is less than 20%: VS + S + MS + mS + NS (NS: Non Suitable).

These five levels are provided for rainfed agriculture and for rainfed and irrigated agriculture. In other words, the GAEZ methodology proposes a calculation of the gain from irrigation, in relation to local physical data (topography, soil texture, soil drainage, etc.) which determine the viability of irrigation. However, this calculation does not take into account water availability and quality (see Chapter 2, p. 35).

The quantification of newly-cultivated areas and newly-irrigated lands in 2050 in each region, and thereby of the remaining forest and pasture areas, is based on:

Table 6.1 Land use and variation rates: 1961–2000 and 2000–2050 in Agrimonde 1, Agrimonde GO and global orchestration

Region	Type of land use	Areas in Agribiom (million ha)		Rate of variation 1961-2000 (%)	Rate of variation 2000-2050 (%)		
		1961	2000		Agri-monde 1	Agri-monde GO	Global Orchestra-tion
MENA	Crops (annual and perennial)	73	83	+14	+8	+12	+2
	Incl. Irrigated land	11	22	+98	0	NA	+7
	Pastures	235	327	+39	−2	−2	+11
	Forest	49	33	−33	0	−35	−50
SSA	Crops (annual and perennial)	144	192	+33	+76	+58	+61
	Incl. Irrigated land	3	7	+100	+105	NA	+17
	Pastures	767	782	+2	−12	+49	+31
	Forest	707	637	−10	−9	−31	−45
LAM	Crops (annual and perennial)	102	162	+58	+91	+64	+54
	Incl. Irrigated land	8	19	+125	+11	NA	+33
	Pastures	462	555	+20	−20	−1	−10
	Forest	1,030	937	−9	−4	1	−1
ASIA	Crops (annual and perennial)	369	455	+23	+23	+11	+9
	Incl. Irrigated land	76	154	+103	+6	NA	+8
	Pastures	416	565	+36	−9	+30	+13
	Forêt	526	497	−5	−10	−11	−16
FSU	Crops (annual and perennial)	240	203	−15	+53	+10	+1
	Incl. Irrigated land	9	21	+123	0	NA	+9
	Pastures	302	359	+19	−16	−41	−35
	Forest	913	843	−8	0	+12	+9
OECD	Crops (annual and perennial)	426	418	−2	+18	+11	+10
	Incl. Irrigated land	27	47	+72	−1	NA	+10
	Pastures	817	752	−8	−23	−19	−19
	Forest	1,071	978	−9	+10	+13	+10
World	Crops (annual and perennial)	1,354	1,513	+12	+39	+23	+19
	Incl. Irrigated land	135	269	+100	+6	NA	+10
	Pastures	2,999	3,340	+11	−15	+7	+2
	Forest	4,296	3,925	−9	−1	−1	−1

- the cultivation potential for rainfed and irrigated crops, as proposed in the FAO and by Fischer et al. (2001, 2000, 2002),

- past dynamics of land use[2],
- assumptions proposed in other global agriculture scenarios: the scenarios of the MA up to 2050 (MA 2005b), the DGR scenario of Michel Griffon up to 2050 (Griffon 2006), the scenarios up to 2025 from the study "World Water and Food to 2025: Dealing with Scarcity" which results from a partnership between IFPRI and the International Water Management Institute (IWMI): "Business-As-Usual" (BAU), "Water Crisis" (CRI), "Sustainable Water Use" (SUS) (Rosegrant et al. 2002),
- factors liable to promote the extension or reduction of agricultural areas, such as the impact of climate change on the cultivation potential, various social pressures which could promote biodiversity conservation, etc.

Land conversion concerns both forests and pastures. For each region, a rule whereby the distribution between the proportions of land taken from forests and from pastures was established, limiting the environmental impacts of the extension of agricultural land. In particular, we sought to limit deforestation, even when this implied sharp divergence from current trends.

The specific quantification of areas devoted to the production of biofuel crops in each region relied on an analysis of the different forms of bio-energy production that could be produced in 2050 on non-food-producing lands. We made the assumption that a new generation of biofuels manufactured from ligno-cellulose or food by-products (e.g. cooking oil or other household refuse), or even from micro-algae with a high lipid yield, would have emerged by 2050. The advantage would be to shift the energy production effort to areas not substituted for food crop areas, especially to forests, by taking advantage of their regeneration capacity. In Agrimonde 1, areas occupied by energy crops or plantations account for less than 10% of the total cultivated land in most regions of the world. After analysing current trends, we made the assumption that this threshold would be exceeded for Latin America (LAM) and OECD-1990 (OECD) where areas with energy crops or plantations account for as much as 20% of cultivated land in 2050.

Land Use in Middle East—North Africa in 2050

FAO data show that between 1961 and 2000, cultivated areas in Middle East—North Africa (MENA) grew by 0.33% per annum, irrigated areas by 1.77% per annum (i.e. a 100% increase for the entire period), and pastures by 0.86% per annum. Meanwhile, forests declined by 1% per annum.

In 2000, cultivated areas in MENA totalled 83 million ha, that is, almost all the land whose yield potential was 20% greater than the maximum attainable yield

[2] In the Agrimonde scenarios, six types of area are distinguished: areas with crops and plantations including irrigated land and areas devoted to non-food biomass production, pastures, forests and immerged areas and other (including urban areas, infrastructures, mountains). For more details, see Chap. 2.

(VS+S+MS+mS) in rainfed and irrigated agriculture (Fischer et al. 2001, 2000, 2002) (Fig. A2.4). FAO data indicate that the total amount of land suitable for agriculture (crop and permanent crop) is around 99 million ha, which means that in 2000, 92% of this area was already under crops (FAO 2003a). "In a few countries the land balance is negative—that is, more land is being cropped than is suitable for rainfed cropping. This is possible where, for example, land that is too sloping or too dry for rainfed crops has been brought into production by terracing or irrigation." (FAO 2003a). It therefore seems that the reserves of cultivable land have already been exhausted in MENA, and that the remaining potential depends essentially on irrigation.

Given the quasi-saturation of the cultivation potential in MENA, and the fact that this potential could decrease further due to climate change[3], the Agrimonde 1 scenario foresees a cultivated area of 90 million ha in 2050 in MENA, with a negligible area devoted to biofuel crops. This means that there are only 7 million ha of newly-cultivated land compared to the year 2000 (a growth of cultivated areas of+0.16% per annum or 8% in 50 years), based on currently unexploited areas suitable for rainfed agriculture in semi-arid zones. This assumption implies the cultivation of all land with a yield potential greater than 20% of the maximum attainable yield (VS+S+MS+mS). It remains only very slightly less than the low variant of the MA scenarios (96 million ha in Global Orchestration) and to that considered in Michel Griffon's DGR scenario (99 million ha excluding biofuel crops) (Figs. 6.1 and 6.2). Although modest, this increase is close to a continuation of the trends observed between 1961 and 2000 (+0.33% per annum), which would lead to the cultivation of 98 million ha in 2050. The 7 million ha newly-cultivated between 2000 and 2050 in Agrimonde 1 have been taken from pastures, given that forests must be preserved to protect water resources. This implies a switch in the growth trend of pastures (−0.04% per annum between 2000 and 2050, against +0.86% per annum between 1961 and 2000), and a stop to the deforestation observed between 1961 and 2000.

In MENA, 53% of renewable water resources were already mobilised by irrigation in 1997–1999 (FAO 2003a). This region has therefore exceeded the threshold of 40%, beyond which water availability is considered a critical factor[4] (FAO 2002). Given the fact that in this region the pressure on water sources is already considerable and the efficiency of irrigation relatively high [40% (FAO 2003a)], and that climate change could exacerbate water stress, the Agrimonde 1 scenario has not envisaged an increase in the overall area under irrigation. This assumption is also found in the MA scenarios where irrigated areas increase from 10 to 11 million ha between 2000 and 2050, in the highest variant (Global Orchestration) (Fig. 6.3), and in the IFPRI-IWMI scenarios for 2025 where irrigated areas under cereal crops

[3] In its 4th Report, the IPCC Panel 2 foresees a temperature increase of 2 to 3.7 °C in MENA by 2050, depending on the season and the scenario considered (A1FI and B1). Variations in rainfall are wider in B1: −9% in March-April-May to+29% in September-October-November (IPCC 2007b).

[4] Libya and Saudi Arabia use more water for irrigation annually than their renewable resources supply, by drawing on fossil reserves of underground water (FAO 2003a).

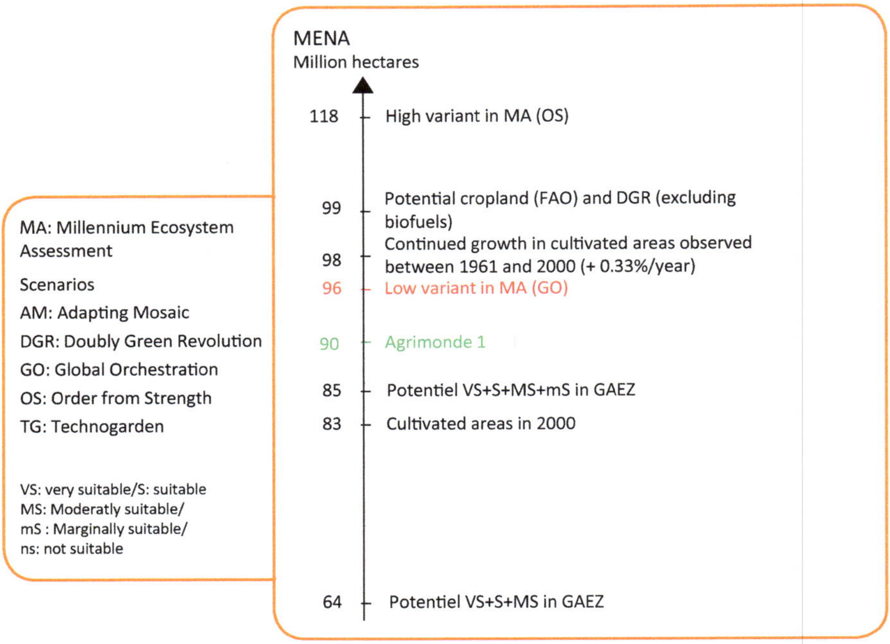

Fig. 6.1 MENA: References for the quantification of cultivated areas in 2050 in the Agrimonde 1 scenario

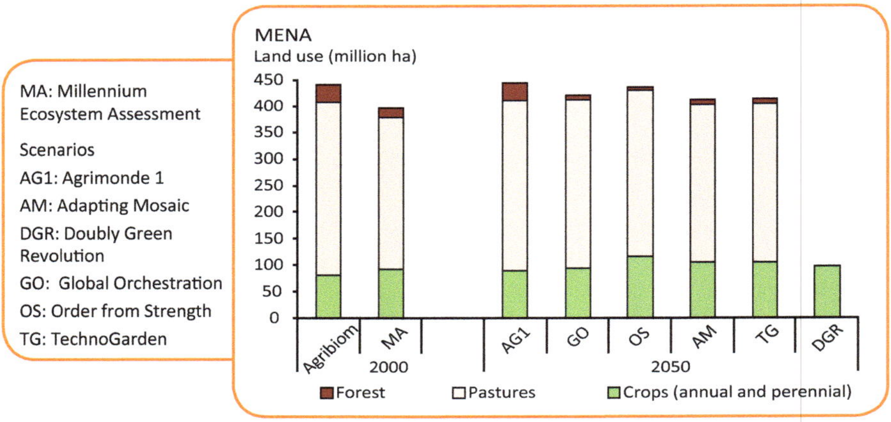

Fig. 6.2 MENA: Land use in 2000 and in 2050 in the Agrimonde 1, Doubly Green Revolution, and MA scenarios

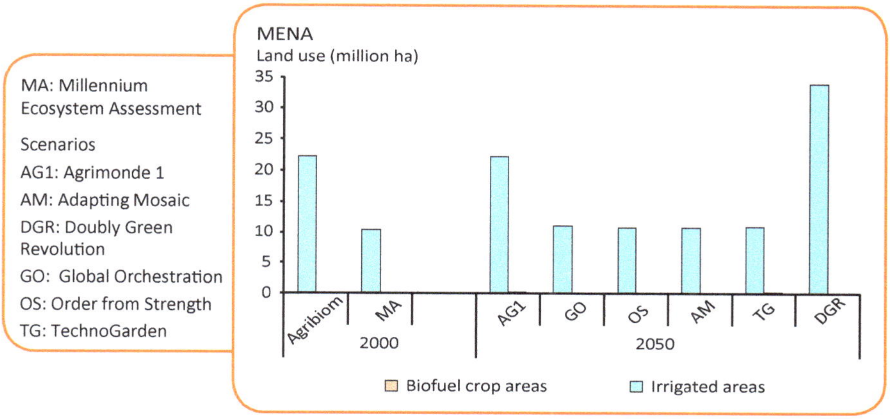

Fig. 6.3 MENA: Irrigated areas and biofuel crop areas in 2000 and in 2050 in the Agrimonde 1, Doubly Green Revolution, and MA scenarios

increase from 9.8 to 10.8 million ha between 1995 and 2025 in the highest variant (BAU)[5].

Finally, in a context in which agricultural land devoted to food production is already limited, it did not seem coherent in the Agrimonde 1 scenario to foresee a significant proportion of the land devoted to biofuel crops. In the MA scenarios the maximum is 0.2 million ha in 2050 (in this count, in addition to areas under biofuel crops, the MA includes areas devoted to wood energy).

Land Use in Sub-Saharan Africa in 2050

FAO data show that between 1961 and 2000, cultivated areas in sub- Saharan Africa (SSA) grew by 0.74 % per annum, irrigated areas by 1.79 % per annum (almost a 100 % increase for that period) and pastures by 0.04 % per annum. In parallel, forests decreased by 0.26 % per annum.

In 2000, some 190 million ha were cultivated in SSA. Fischer et al. (2001, 2000, 2002) assess the cultivation potential for a yield that is greater than 80 % of the maximum attainable yield (VS) of rainfed and irrigated agriculture, at 426 million ha, which is more than double the area effectively cultivated in 2000 (Fig. A2.4). For a yield greater than 60 % of the maximum attainable yield (VS + S), it is assessed at close to 800 million ha, which is four times the area cultivated in 2000. Finally, according to FAO data, land suitable for agriculture (crops and permanent crops)

[5] In the IFPRI-IWMI scenarios, annual growth rates of irrigated areas with cereal crops in MENA range from + 0.10 % in the CRI scenario, to + 0.17 % in the SUS scenario, and + 0.32 % in the BAU scenario, compared to a reference of 9.8 million ha in 1995 (Rosegrant et al. 2002).

totals 1,031 million ha, which is over five times the area actually cultivated in 2000 (FAO 2003a).

These figures show how large the reserves of agricultural land are in this region. Various scenarios in terms of land use can therefore be considered. A scenario maintaining the same growth rate of cultivated lands as that observed between 1961 and 2000 would assume a cultivated area of 278 million ha in 2050. The MA scenarios foresee a slightly larger cultivated area of between 291 and 337 million ha in 2050. This moderate growth of cultivated areas is counterbalanced, in the MA scenarios, by the growth of other factors of production gains such as yields. In his DGR scenario, Michel Griffon assumes the opposite. Considering that the capacity of SSA to increase its future agricultural yields is weak, he foresees a fourfold increase in cultivated areas by 2050. The total would then be around 840 million ha, excluding biofuel crops. In the GAEZ study typology, this figure is midway between the conversion of land with a yield potential greater than 60 % of the maximum attainable yield (VS + S), and that of land with a yield potential greater than 40 % of the maximum attainable yield (VS + S + MS).

In the Agrimonde 1 scenario in SSA, the cultivated area in 2050 is assumed to amount to about 340 million ha (+ 1.14 % per annum between 2000 and 2050), of which almost 40 million are devoted to biofuel crops. This area corresponds to the cultivation of 80 % of the land with a yield potential greater than 80 % of the maximum attainable yield (VS), and than the highest variant of the MA scenarios (337 million ha in TechnoGarden) (Figs. 6.4 and 6.5). As the conservation of forests is a priority in Agrimonde 1, we assumed that 60 % of the newly-cultivated areas were taken from pastures and 40 % from forests. Compared to the trend observed between 1961 and 2000, this implies a reversal of the growth of pastures (−0.24 % per annum between 2000 and 2050), and a reduction of the annual deforestation rate (+0.19 % per annum between 2000 and 2050).

The figures on irrigated areas in SSA vary widely, from one source to another. The IFPRI-IWMI estimates the irrigated area with cereal crops in this region at only 3.3 million ha in 1995, but foresees an annual growth rate of 1.11 % for these areas in the SUS scenario[6]. A rate such as this would make it possible to almost double the irrigated areas in 50 years. The MA, on the other hand, starts with 13.2 million ha irrigated in 2000 and foresees a growth of only 0.27 to 0.30 % per annum up to 2050 (Fig. 6.6). Our assumptions were based on FAO data, estimating that irrigated areas covered 6.75 million ha in 2000 in SSA (included in the 190 million ha cultivated). This is double the area recorded for 1961. We chose to repeat this 100 % increase for the period up to 2050, to attain an irrigated area of 14 million ha, which is consistent with the areas envisaged in the MA scenarios (between 15.1 and 15.5 million ha), and a growth rate close to the one foreseen in the IFPRI-IWMI BAU scenario for 2025 (+ 1.47 per annum in Agrimonde 1, and + 1.33 % per annum in the IFRI-IWMI BAU scenario). This assumption demonstrates a desire to make best use of the

[6] In the IFPRI-IWMI scenarios, annual growth rates of irrigated areas with cereal crops in SSA range from + 1.11 % in the SUS scenario to + 1.26 % in the CRI scenario and + 1.33 % in the BAU scenario (Rosegrant et al. 2002).

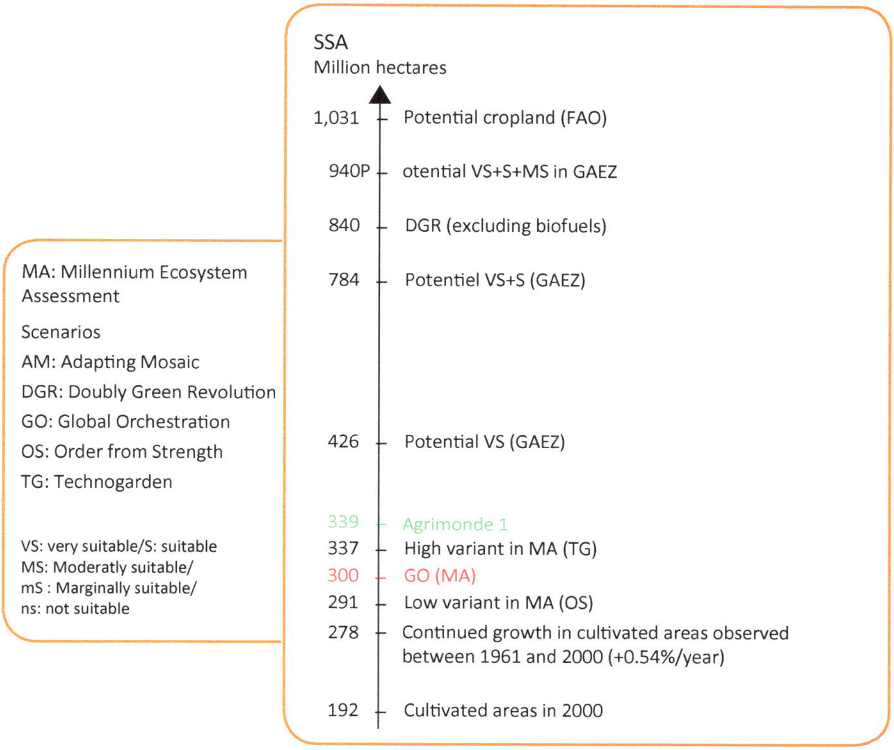

Fig. 6.4 SSA: References for the quantification of cultivated areas in 2050 in the Agrimonde 1 scenario

region's water resources, given that only 2% of the renewable water available for irrigation is used in SSA today (FAO 2003a).

Finally, in the Agrimonde 1 scenario, in the period from 2000 to 2050, areas devoted to biofuels in SSA have grown in the same proportion as in the MA Global Orchestration and Adapting Mosaic scenarios (39 million ha, in this count, in addition to areas under biofuel crops, the MA includes areas devoted to wood energy) (Fig. 6.6). This gives an average growth rate for the MA scenarios, in which these areas range from 3 million ha in Order from Strength, to 65 million ha in TechnoGarden. In Agrimonde 1 the areas devoted to biofuel crops serve primarily to satisfy local demand. Some industrial plants producing second-generation biofuels could have been set up, mainly with foreign capital, to exploit biomass in the Congo Basin and plantations in South Africa.

To recap, in the Agrimonde 1 scenario the assumptions on cultivated areas in SSA remain moderate compared to the large reserve of land suitable for crops in this region. They are consistent with the IPCC's forecast of aridification of the Sahel and

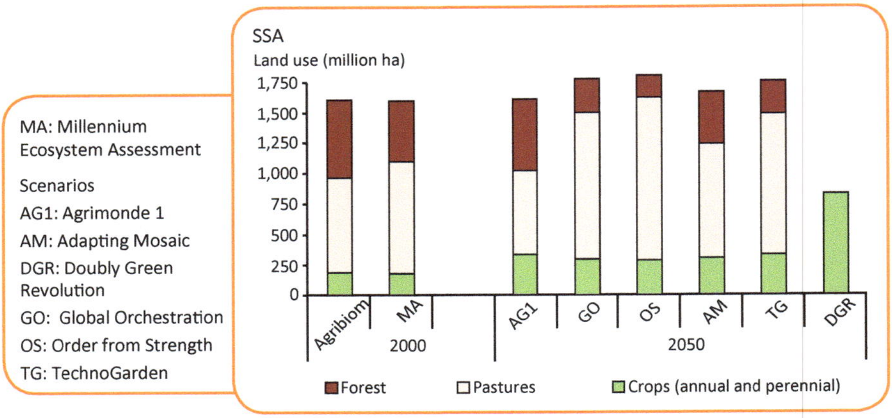

Fig. 6.5. SSA: Land use in 2000 and in 2050 in the Agrimonde 1, Doubly Green Revolution, and the MA scenarios

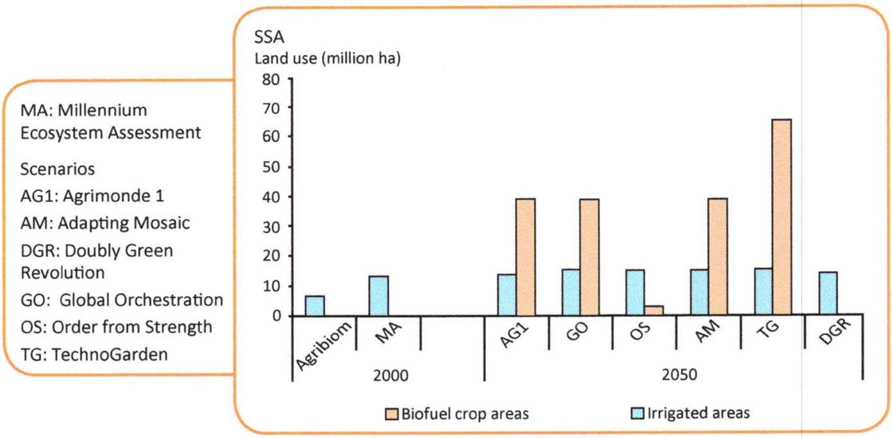

Fig. 6.6 SSA: Irrigated areas and areas devoted to biofuel crops in 2000 and in 2050 in the Agrimonde 1, Doubly Green Revolution, and MA scenarios

southern Africa[7], which could reduce the cultivation potential thus far estimated. Our assumptions suppose that cultivated areas have encroached on the forests of the Congo Basin, primarily in the East, and on most of the fallow lands and pastures of West Africa. The implications of this loss of natural and semi-natural areas are discussed in Chap. 9.

[7] The fourth IPCC report does not give projections of changes in temperature and rainfall in 2050. It has however estimated that temperatures could rise by 5 °C in the west of the Sahel and in South Africa up to 2100. Precipitation is likely to become far more erratic, increasing by up to 20 % in the area of Sudan-Ethiopia in the rainy season, and decreasing by as much in the dry season (IPCC 2007c).

Land Use in Latin America in 2050

FAO data show that between 1961 and 2000, cultivated areas in LAM increased by 1.19 % per annum, irrigated areas by 2.11 % per annum (i.e. + 125 % between 1961 and 2000) and pastures by 0.47 % per annum. In parallel, forest areas decreased by 0.24 % per annum.

In 2000, some 160 million ha were cultivated in LAM, and the reserves of cultivable land were still considerable. The cultivation potential for a yield greater than 80 % of the maximum attainable yield (VS) in rainfed and irrigated agriculture is thought to be more than double that, according to the GAEZ study (Fischer et al. 2000, 2001, 2002) (Fig. A2.4), and the total amount of land suitable for agriculture (crops and permanent crops) is evaluated by the FAO at six and a half times more (FAO 2003a).

As in SSA, such unexploited agricultural potential allows for much latitude in scenario-building. Yet the different scenarios make relatively similar assumptions on the extent to which cultivated areas will grow, ranging from one and a half to two and a half times the area recorded for 2000, that is, 265 million ha in Global Orchestration and 427 million ha in Michel Griffon's highest variant of the DGR scenario[8] (excluding areas with biofuel crops, in the latter scenario).

In the Agrimonde 1 scenario, cultivated areas in LAM have almost doubled between 2000 and 2050, up to 310 million ha (of which 60 million are devoted to biofuel crops). This figure is at the top end of the range given by the MA scenarios (between 265 and 325 million ha), and matches the low variant given by Michel Griffon in his DGR scenario (308 million ha, not including areas with biofuel crops) (Figs. 6.7 and 6.8). In terms of trends, this assumption implies a growth rate of cultivated land that is slightly higher than that observed between 1961 and 2000 (+ 1.30 % between 2000 and 2050, compared to + 1.19 % between 1961 and 2000). Furthermore, we assumed that in Agrimonde 1 there has been a forest preservation effort in LAM between 2000 and 2050. Three-quarters of newly-cultivated areas have therefore been taken from pastures and only one quarter from forests. This implies a very significant reversal of the growth trend of pastures (− 0.44 % per annum between 2000 and 2050, compared to + 0.47 % per annum between 1961 and 2000), and a drastic reduction in the rate of deforestation (− 0.24 % per annum between 1961 and 2000, compared to − 0.08 % per annum between 2000 and 2050).

Only 1 % of the renewable water resources is now used for irrigation in LAM (FAO 2003a). Agriculture is essentially rainfed, except in some arid areas and on the west coast. The potential for rainfed crops is considerable and very close to the potential for rainfed and irrigated crops (320 million ha for a yield potential of over 80 % of the maximum attainable yield in rainfed crops (VS), according to Fischer et al. (2001, 2000, 2002), compared to 344 million ha for rainfed and irrigated crops). We therefore consider that between 2000 and 2050, irrigation has not been the main way of ensuring the extension of cultivated land foreseen in Agrimonde

[8] Michel Griffon proposes a high variant and a low variant for cultivated areas in LAM, as opposed to a single variant for the other regions (Griffon 2006).

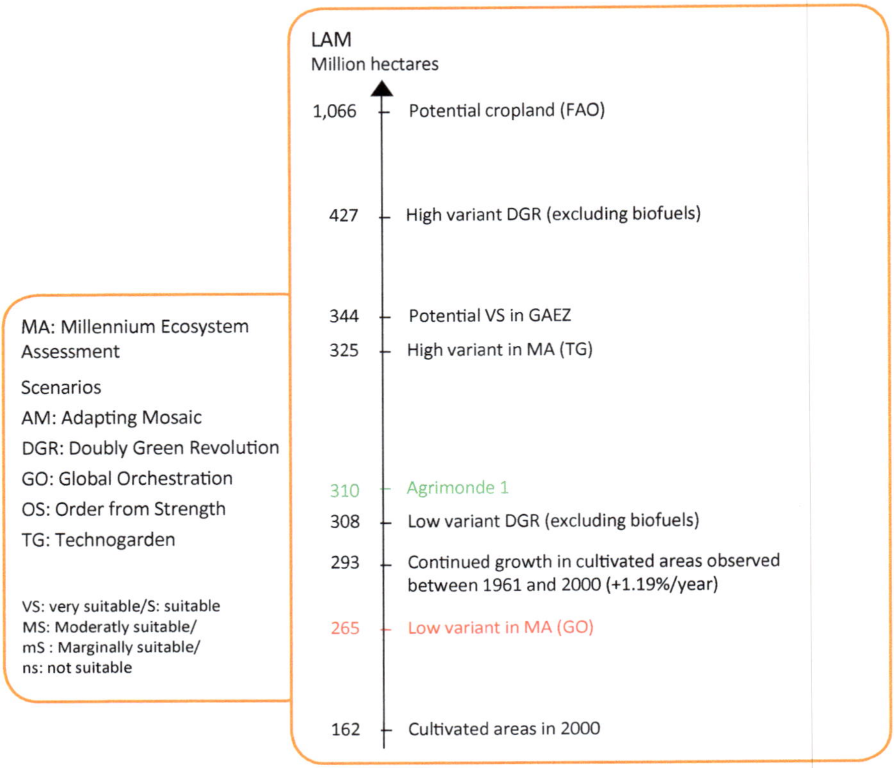

Fig. 6.7 LAM: References for the quantification of cultivated areas in 2050 in the Agrimonde 1 scenario

1, and that irrigated areas have remained at much the same level as in 2000. This assumption is close to that put forward in the other scenarios (+3 million ha in 50 years in the MA scenarios, +2 million ha in 30 years for cereal crops in the IFPRI-IWMI scenarios[9]), even though the irrigated areas in the reference year vary widely from one study to another. They are estimated at 19 million ha in 2000 in Agrimonde (based on FAO data), at only 9 million in 2000 in the MA scenarios, and at 7.5 million for irrigated areas with cereal crops in 1995 in the IFPRI-IWMI scenarios (Rosegrant et al. 2002).

In LAM the biofuel industry has grown rapidly in the late twentieth and early twenty-first century, especially in the sugarcane and soya supply chains. We consider that the amounts invested, the technical capacities deployed and the vast availability of cultivable land are all incentives to pursue its development. In Agrimonde 1, the areas devoted to the production of biofuels have therefore continued to grow.

[9] In the IFPRI-IWMI scenarios, annual growth rates of irrigated areas of cereal crops in LAM range from +0.76 % in the CRI scenario, to +0.86 % in the SUS scenario, and +0.90 % in the BAU scenario (Rosegrant et al. 2002).

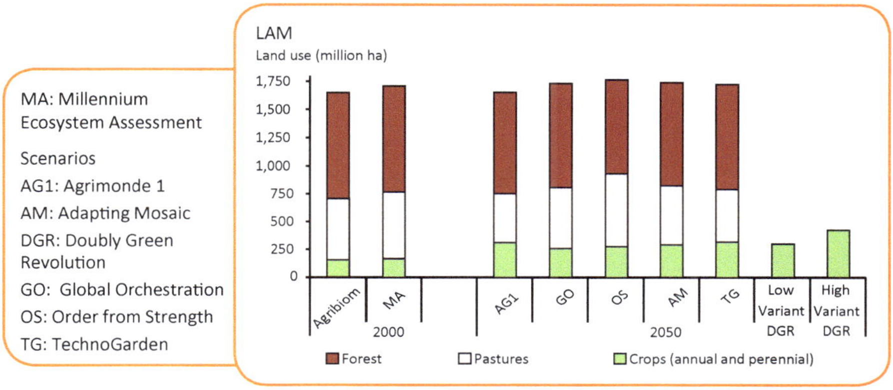

Fig. 6.8 LAM: Land use in 2000 and in 2050 in the Agrimonde 1, Doubly Green Revolution, and MA scenarios

Initially they were used for regional production of first-generation biofuels. Then, shortly before 2050, heavier industries producing second-generation biofuels from lignocellulose were set up. The assumption is made that certain countries like Brazil, which are rich in fresh and sea water, have developed their capacities to produce biofuels from seaweed. With 60 million ha devoted to energy crops (i.e. 19% of the cultivated land), the area considered in Agrimonde 1 is situated within the same range as the assumptions in the MA scenarios (between 47 and 93 million ha, in this count, in addition to areas under biofuel crops, the MA includes areas devoted to wood energy) and more specifically, at the same level as in the Adapting Mosaic scenario (Fig. 6.9).

To sum up, the agricultural potential of LAM is unquestionable but the Agrimonde 1 scenario assumes a moderate growth of cultivated areas up to 2050, allowing for the preservation of forests. A substantial proportion of the *Cerrado* or pastures have been used for crops. The farming systems adopted in these areas and the incentives to preserve forests are described in Chap. 9.

Land Use in Asia in 2050

FAO data for the period from 1961 to 2000 show that cultivated areas in Asia (ASIA) grew by 0.54% per annum, irrigated areas by 1.84% per annum (roughly a 100% increase for this period) and pastures by 0.79% per annum. Meanwhile, forests shrank by 0.14% per annum.

In ASIA, the cultivation potential for a yield greater than 60% of the maximum attainable yield (VS+S) in rainfed and irrigated agriculture is 407 million ha, according to Fischer et al. (2001, 2000, 2002) (Fig. A2.4). This area was exceeded in 1985 and the cultivated area is now around 450 million ha, which means that Asian farmers are already using land with a yield potential of less than 60% of the

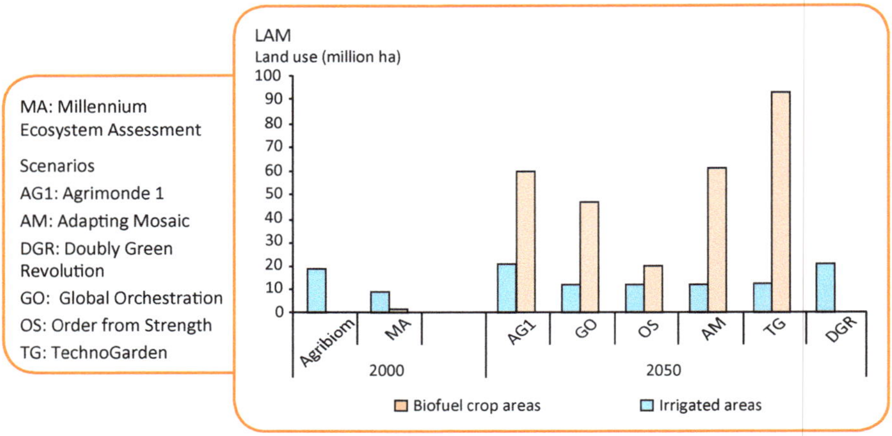

Fig. 6.9 LAM: Irrigated areas and areas devoted to biofuel crops in 2000 and in 2050 in the Agrimonde 1, Doubly Green Revolution, and the MA scenarios

maximum attainable yield. According to FAO data, the amount of land suitable for agriculture (crop and permanent crop) in ASIA is closer to 586 million ha, of which 63 % is already exploited in East Asia and 94 % in South Asia (FAO 2003a). Most scenarios on cultivated areas in ASIA take into account the saturation of the cultivation potential in this region.

In the Agrimonde 1 scenario the cultivated area in ASIA in 2050 represents 560 million ha, 20 million of which are devoted to biofuel crops. This figure implies an annual growth rate of 0.42 % between 2000 and 2050, which is consistent with the highest variant in the MA scenarios, observed in Order from Strength, and that of Michel Griffon's DGR scenario (excluding areas under biofuel crops) (Figs. 6.10 and 6.11). Moreover, new cultivated lands are assumed to have been taken from pastures and forests in equal proportions. This implies a reversal of the current growth of pastures (−0.20 % per annum between 2000 and 2050) and a very slight increase in the annual deforestation rate (0.22 % per annum) compared to that of the period from 1961 to 2000 (0.14 % per annum).

Projects to develop irrigation in this region and especially in China involve the construction of dams whose social and environmental effects are highly controversial issues. The Agrimonde 1 scenario therefore assumes a very weak growth of irrigated areas (+0.11 % between 2000 and 2050, compared to +1.84 % between 1961 and 2000), to reach a total of 163 million ha under irrigation in 2050 (included in the 560 million ha cultivated). This figure may seem low compared to FAO projections (+0.33 % per annum up to 2030) which assume that East Asia will exploit three-quarters of its irrigable land in 2030, and South Asia almost 90 % (FAO 2003a). It is also lower than the projection of the IFPRI-IWMI SUS scenario for 2025, (+0.14 % per annum for cereal crops up to 2025), based on a sustainable development logic[10].

[10] In the IFPRI-IWMI scenarios the annual growth of irrigated areas under cereal crops in ASIA range from+0.03 % in the CRI scenario,+0.23 % in the SUS scenario, and 0.32 % in the BAU

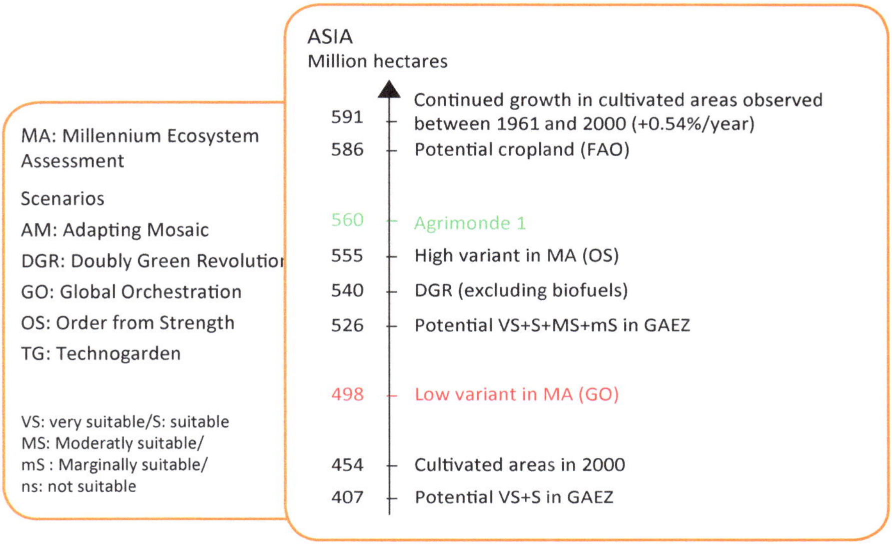

Fig. 6.10 ASIA: References for the quantification of cultivated areas in 2050 in the Agrimonde 1 scenario

This assumption of the Agrimonde 1 scenario consists in particular in considering that available water resources have been a factor limiting the development of irrigation. It is situated in the upper range of the MA scenarios (growth between −0.04% in Global Orchestration and +0.15% in Order from Strength) (Fig. 6.12).

Finally, in Agrimonde 1 it is assumed that intense pressure on food-producing land has limited the development of biofuels between 2000 and 2050[11]. In ASIA, the area devoted to biofuels covers 20 million ha in 2050 and accounts for less than 4% of all cultivated land. This assumption is close to that of TechnoGarden (in this count, in addition to areas under biofuel crops, the MA includes areas devoted to wood energy) (Fig. 6.12). In Agrimonde 1, biofuels are produced primarily from oil palms in South-East Asia and Indonesia. The main source of biofuels in ASIA in 2050 may however be seaweed, cultivated in marine areas. This assumption would confirm the growth of seaweed crops in recent years: from 1996 to 2006 the production of red seaweed almost doubled (FAOSTAT 2008), and 95% of this production was in ASIA (Kaas 2006).

This choice of assumptions on cultivated areas, irrigated areas, pastures and forests is highly space-consuming. It is largely motivated by the idea that, between 2000 and 2050, the growing farming population in this region has tried to push back the frontiers of agricultural areas to perpetuate its activity[12]. This will result in a sig-

scenario (Rosegrant et al. 2002).

[11] This assumption is also made by Johnson et al. (2008).

[12] The cultivated area per inhabitant will thus be reduced to 0.23 ha in 2050, compared to 0.31 ha in 2000 and 0.51 in 1961.

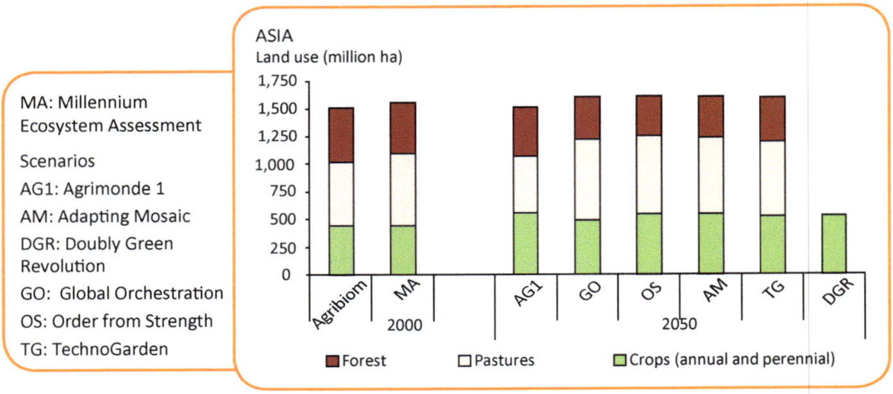

Fig. 6.11 ASIA: Land use in 2000 and in 2050 in the Agrimonde 1, Doubly Green Revolution, and MA scenarios

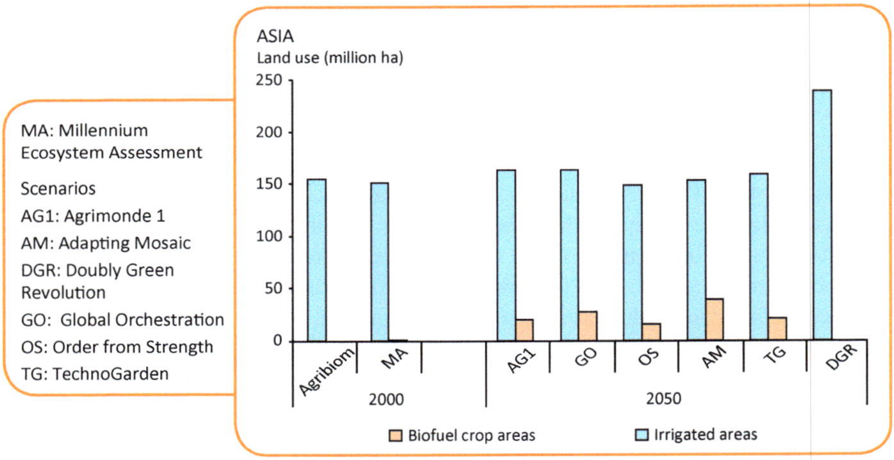

Fig. 6.12 ASIA: Irrigated areas and areas devoted to biofuel crops in 2000 and in 2050 in the Agrimonde 1, Doubly Green Revolution, and MA scenarios

nificant advancement of the pioneering fronts in Indonesia and the north of China (Griffon 2006) and in the shifting of production areas northwards (Taithe 2007) where the climate will become more favourable to agriculture than in South Asia[13]. For this assumption on the growth of cultivated areas to be possible and compatible

[13] In its fourth Report, the IPCC Panel 2 foresees temperature increases of 3.1 to 6.6 °C in North Asia by 2050, depending on the season and the scenario considered (A1FI and B1). Global rainfall also increases (++N in June-July-August in B1 to +35 % in December-January-February in A1FI). Thus, the climate could become more favourable to agriculture. However in South Asia, the hot and humid climate will intensify, with temperature increases of 0.9 to 3.1 °C, and rainfall increases of 0 to 26 %, depending on the season and the scenario considered (A1FI and B1) (IPCC 2007b).

with sustainability objectives, it goes hand-in-hand with major changes in farming practices specified in Chapter 9.

Land Use in the Former Soviet Union in 2050

FAO data show that between 1961 and 2000, cultivated areas in the former Soviet Union (FSU) decreased by 0.43 % per annum, irrigated areas increased by 2.09 % per annum (+ 125 % between 1961 and 2000) and pastures increased by 0.44 % per annum. Meanwhile, forests decreased by 0.20 % per annum. Data series for the past are however highly chaotic and their reliability is doubtful.

In 2000, all the land of FSU with a yield potential greater then 80 % of the maximum attainable yield (VS) in rainfed and irrigated crops was already cultivated. There were 203 million ha under crops, which means that there remained only 13 % of land with a yield potential greater than 60 % of the maximum attainable yield (VS+S), or 40 % of land with a yield potential greater than 40 % (VS+S+MS) (Fischer et al. 2001, 2000, 2002) (Fig. A2.4). Theoretically, this region could therefore still extend its cultivated areas but only moderately. The effects of climate change could however free new reserves of agricultural land in the future (Box 6.2).

In the Agrimonde 1 scenario, the cultivated area in FSU is estimated at 310 million ha in 2050, of which 10 million are devoted to biofuels. This figure is equivalent to the highest variant in the MA scenarios (304 million ha in TechnoGarden) but remains lower than the GAEZ study's 339 million ha of cultivable land with a yield superior to 40 % of the maximum attainable yield (VS+S+MS), without taking the effects of climate change into account (Fischer et al. 2000, 2001, 2002) (Figs. 6.13 and 6.14). The current cultivation potential allows for such an increase in cultivated areas. However, in order to preserve this region's forests, it is assumed that cultivated lands will be extended by 107 million ha in the south of the region on what is currently pastureland. The pastures will shift northwards, primarily onto 48 million ha of land made available by the melting of the permafrost. The forests have thus been preserved. This implies a stagnation of forest areas between 2000 and 2050, whereas in all the MA scenarios they have grown in this region. Pastures have shrunk by 0.36 % per annum between 2000 and 2050, i.e. twice as slowly as in the MA scenarios, thus putting an end to the growth of pastures observed between 1961 and 2000 (0.44 % per annum). New land for crops has thus been taken primarily from pastures in the south of FSU, and has pushed the pastures further north as the permafrost melts.

In Agrimonde 1, irrigated areas in 2050 have remained at their 2000 level due to the scarcity of water in this region, especially in Central Asia. This is consistent with the assumptions of the MA scenarios which have foreseen a maximum increase in irrigated areas by 1.5 million ha and a minimum decrease by at least 2.3 million ha (MA 2005b)[14]. Finally, the MA scenarios have all assumed a growth

[14] In the IFPRI-IWMI scenarios, figures on irrigated areas under cereal crops in FSU are included in a broader category called "Developed Countries". This explains why no reference is made to

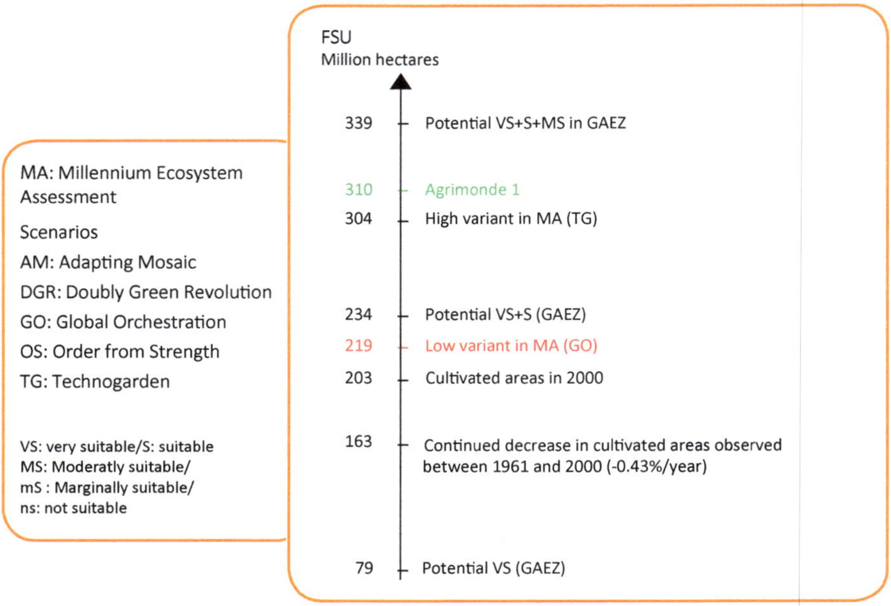

Fig. 6.13 FSU: References for the quantification of cultivated areas in 2050 in the Agrimonde 1 scenario. (In FAO studies and Michel Griffon's book, figures on FSU are included in a broader category, "Industrialised countries". These two sources were therefore not used as references in setting the quantitative assumptions for this region in the Agrimonde 1 scenario)

Box 6.2—Climate change and cultivable potential in the former Soviet Union

In the regions north of the 60th parallel, the melting of the permafrost and the increase in precipitation, both due to rising temperatures and the slowing down of the thermohaline circulation, are expected to make vast plains available for agriculture. An increase by 4 °C could imply a shift in the southern limit of Siberian permafrost 100–200 km northward (IPCC 2001). FAO TERRASTAT data show that arable land in FSU could increase from 216 to 314 million ha. Fischer et al. predict a 64 % expansion of potential land suitable for crop cultivation between 1990 and 2080 in the Russian Federation (increase of over 245 million ha) (Fischer et al. 2005).

of biofuel crops, with the cultivation of 13 to 90 million ha dedicated to this purpose (in this count, in addition to areas under biofuel crops, the MA includes areas devoted to wood energy) (MA 2005b) (Fig. 6.15). In the Agrimonde 1 scenario, because of the potential of the region's forests, it is assumed that biofuels have been developed primarily from forest plantations. Agricultural areas devoted to biofuels

---

these scenarios in the quantitative assumptions of the Agrimonde 1 scenario.

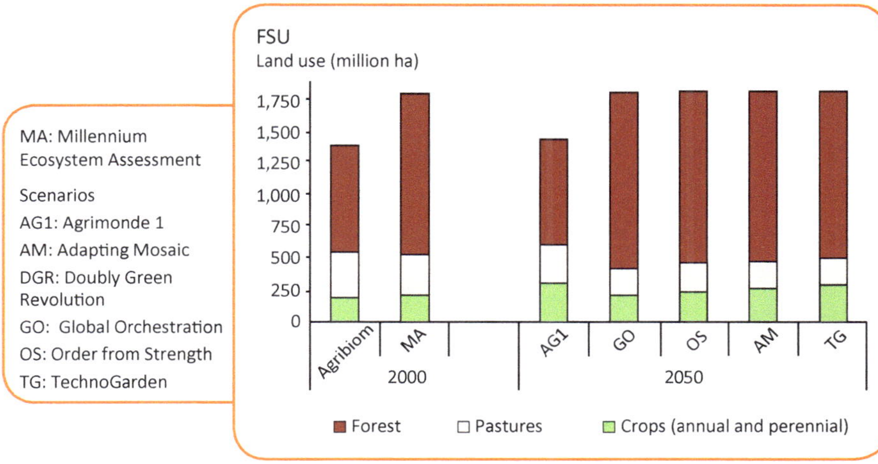

Fig. 6.14 FSU: Land use in 2000 and in 2050 in the Agrimonde 1 and MA scenarios

amount to 10 million ha in 2050 (included in the 310 million ha of cultivated land), a surface area close to that foreseen in Order from Strength.

To sum up, in the Agrimonde 1 scenario there are over 100 million ha of newly-cultivated land in FSU in 2050. This expansion of cultivated areas raises the question of how they are developed and farmed—a question that will be discussed in Chapter 9.

Land Use in the OECD-1990 Region in 2050

FAO data show that between 1961 and 2000, cultivated areas in the OECD region decreased on average by 0.05 % per annum, pastures decreased by 0.21 % per annum, and forests decreased by 0.23 % per annum. Only irrigated areas increased, by 1.42 % per annum.

In 2000 all the land in the OECD region whose yield potential was greater than 80 % of the maximum attainable yield (VS) in rainfed and irrigated agriculture was already cultivated. There were 419 million ha of crops, which, in the typology of the GAEZ study, means that 24 % of the land with a yield potential greater than 60 % of the maximum attainable yield (VS + S) remained available, or 44 % of the land with a yield potential greater than 40 % (VS + S + MS) (Fischer et al. 2001, 2000, 2002) (Fig. A2.4). Theoretically, this region therefore still has reserves of cultivable land that could expand in the future due to the melting of the permafrost in the north of Canada, induced by global warming.

In the Agrimonde 1 scenario, the cultivated area is assumed to reach 495 million ha in 2050 in OECD, of which 95 million are devoted to biofuels. This

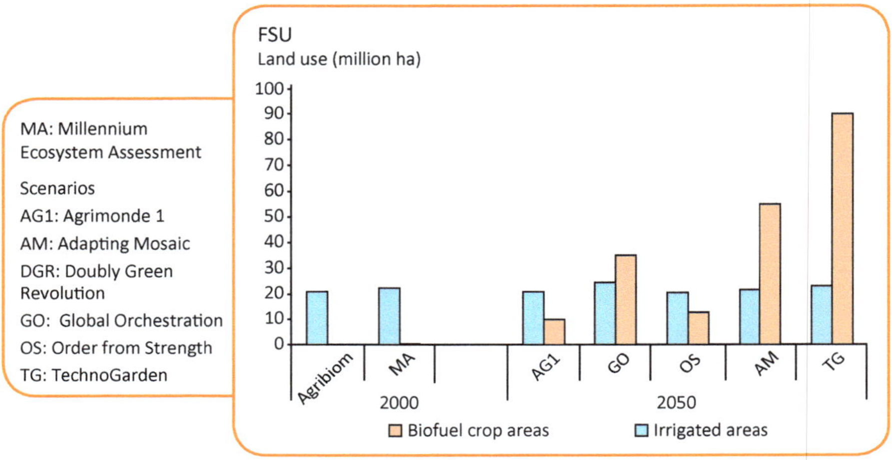

Fig. 6.15. FSU: Irrigated areas and areas devoted to biofuel crops in 2000 and in 2050 in the Agrimonde 1 and MA scenarios

assumption is close to the low variant in the MA scenarios [457 million ha in Global Orchestration (MA 2005b)] (Figs. 6.16 and 6.17). It corresponds to the conversion of 90 % of the lands that the GAEZ study classified as cultivable at a yield level greater than 60 % of the maximum attainable yield (VS + S), without taking the effects of global warming into account (Fischer et al. 2000, 2001, 2002).

Note that in Agrimonde 1 this growth of cultivated areas is due only to the development of energy crops. It even conceals the sharp decrease in food crop areas (− 18 million ha). In parallel, the reforestation observed in the OECD region between 1990 and 2003 has continued and has even been accentuated between 2000 and 2050. Forest areas, which increased by 0.11 % per annum between 1990 and 2003 (+ 14 million ha), have expanded another 0.19 % per annum between 2000 and 2050 in Agrimonde 1. Pastures are the adjustment variable, decreasing by 176 million ha as they are replaced by forests and biofuel crops. Concurrently with their overall decrease in the region, certain pastures nevertheless expand locally in the north of Canada, as the permafrost melts.

As in the MA scenarios, in Agrimonde 1 the expansion of this region's cultivated areas takes place without any significant growth of irrigated lands[15] (Fig. 6.18). The rapid growth of areas devoted to biofuels is equivalent to that foreseen in Adapting Mosaic (in this count, in addition to areas under biofuel crops, the MA includes areas devoted to wood energy), the MA scenario that allocates the most space to biofuel crops [95 million ha as opposed to 62 million ha in TechnoGarden, the low variant in the MA scenarios (MA 2005b)].

[15] In the IFPRI-IWMI scenarios, the figures pertaining to irrigated areas under cereal crops in the OECD region are included in a larger category "Developed Countries". This explains why they are not referred to in the quantitative assumptions for the Agrimonde 1 scenario in OECD.

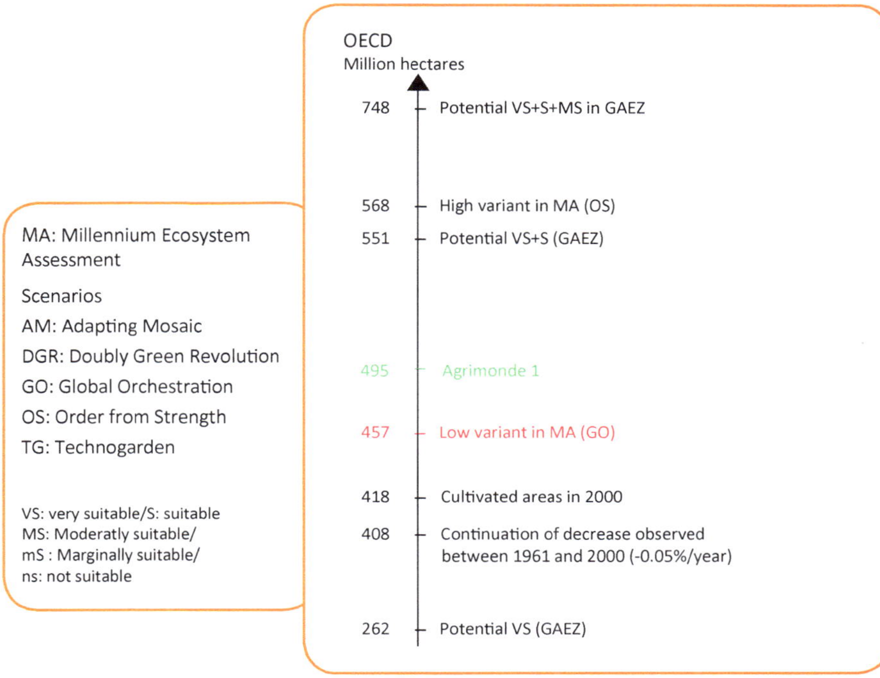

Fig. 6.16 OECD: References for the quantification of cultivated areas in 2050 in the Agrimonde 1 scenario (In FAO studies and Michel Griffon's book, figures on the OECD region are included in a broader category, "industrialised countries". These two sources were therefore not used as references in setting the quantitative assumptions for this region in the Agrimonde 1 scenario)

Land Use in the World in 2050

FAO data show that the world's cultivated areas (arable land and permanent crops) were estimated to be 1.5 billion ha at the beginning of the twenty-first century, i.e. 11 % of the earth's land. This is 12 % higher than the figure recorded in 1961, and represents a mean annual increase of close to 4 million ha over the past 40 years.

This past increase in cultivated areas is distributed differently, depending on the region (Table 6.1). SSA and LAM are the regions in which there has been the most newly-cultivated land, with increases of 34 % and 58 % respectively. Forest surface areas seem to have been the adjustment variable in SSA, whereas their decrease does not entirely explain the increase in cultivated surfaces and pastures in LAM. The cultivation potential, still high in these regions, partly explains the considerable expansion of cultivated land. In contrast, cultivated areas in ASIA and MENA increased moderately (23 % and 14 %). These slow evolutions are explained by the fact that the expansion of cultivated areas is very close to the limit of the cultivation potential. Only the FSU and OECD regions show decreasing cultivated areas. In these regions the cultivation potential has not yet been attained. The decrease of

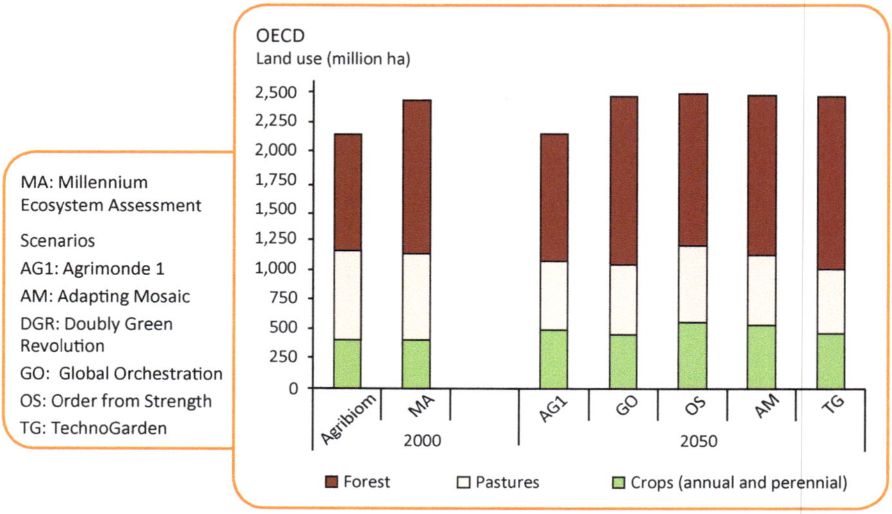

Fig. 6.17 OECD: Land use in 2000 and in 2050 in the Agrimonde 1 and MA scenarios

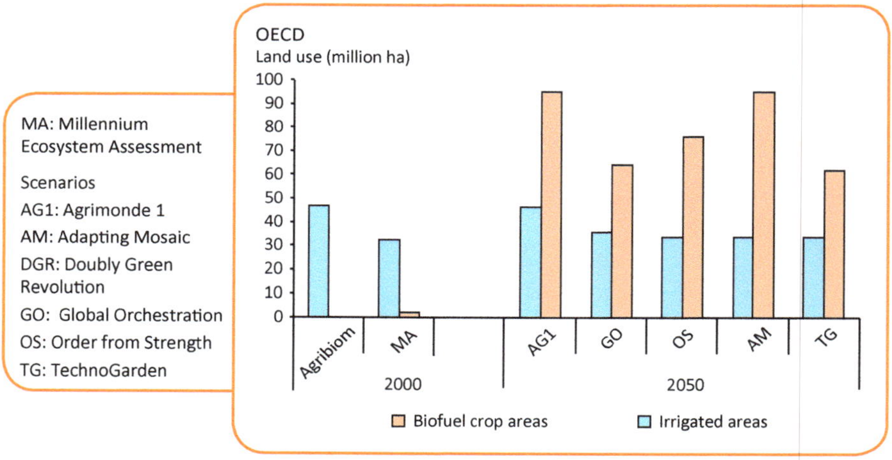

Fig. 6.18 OECD: Irrigated areas and areas devoted to biofuel crops in 2000 and in 2050 in the Agrimonde 1 and MA scenarios

cultivated areas is due to various phenomena: economic slump in FSU and expansion of forests and urbanisation in OECD.

Irrigated areas, which are included in the figures mentioned above, have not followed the same trend. They increased in all the regions by about 100 % between 1961 and 2000, and represented 18 % of the world's cultivated land in 2000. This sharp rise over the past 40 years raises the question of the continuation of this trend, with regard to water resources, especially in those regions where irrigated land already accounts for a large proportion of cultivated areas: ASIA (34 %) and MENA (27 %).

The evolution of pastures between 1961 and 2000 is characterised by two trends distinguishing two groups of regions: those where pastures have rapidly expanded, and those where they remained stable or even decreased. The first group includes LAM (+20 %), ASIA (+36 %), FSU (+19 %) and MENA (+39 %). In the second group we find SSA (stagnation) and OECD where pastures decreased considerably, to the benefit of forests and urbanisation.

In Agrimonde GO, which takes the land use assumptions of Global Orchestration but applies the corrective factor given in Appendix 4 (p. 263) the cultivated land area has increased by 23 % between 2000 and 2050, at an average rate of 6.8 million newly-cultivated hectares per year, i.e. a rate 1.7 times faster than the rate observed between 1961 and 2000 (Fig. 6.19 and Table 6.1). The new cultivated land areas are mainly in SSA and LAM, and to a lesser extent in ASIA and OECD, whereas in the other two regions the cultivated land areas have remained stable. The deforestation and increase of pasture lands observed from 1961 to 2000 have continued in Agrimonde GO, although at a slower rate. Pastures have gained 245 million ha, largely to the detriment of the category "other"[16]. In SSA and in ASIA, the savannah has grown by respectively 380 and 170 million ha, primarily in areas that were previously under forest cover. Trends are the opposite in the OECD and FSU countries, which have replaced respectively some 147 and 144 million ha of pastures essentially by forests between 2000 and 2050. In LAM, deforestation has stopped. Finally, in Agrimonde GO, the total area covered in pastures and crops has increased by 12 % in 2050 compared to 2000.

In the Agrimonde 1 scenario, the world's cultivated areas have grown by 39 % between 2000 and 2050, attaining 2.1 billion ha in 2050, i.e. 15 % of the earth's land surface area. With an average of 12 million ha newly-cultivated annually, the conversion rate has tripled between 2000 and 2050 compared to that of the 1961–2000 period.

The conquest of newly-cultivated land has essentially concerned LAM and SSA which have converted 148 and 147 million ha respectively, followed by FSU and ASIA, which have converted 107 and 106 million ha respectively. The participation of OECD in the phenomenon has been more limited (76 million ha of newly-cultivated land), while the cultivated area of MENA has grown very little (Tables 6.1 and 6.2).

In Agrimonde 1 the preservation of forests is a strong objective. Pastures have become the adjustment variable. They have lost 494 million ha (i.e. 15 % of their surface area) while forests have decreased by only 47 million ha (i.e. 1.2 % of their surface area). The decrease in pasture surface areas is greatest in OECD (− 176 million ha), where a quarter of their total area has been converted to biofuel crops, but

[16] The category 'Other' corresponds to emerged land other than that under crops, pastures or forests.

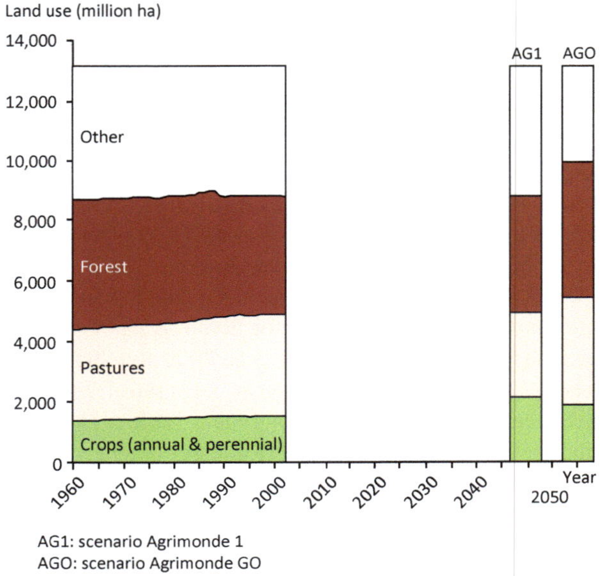

Fig. 6.19 Land use in the world from 1961 to 2003 and in 2050 in the Agrimonde scenarios

has also been substantial in LAM and SSA (−110 and −91 million ha respectively). Meanwhile, efforts to preserve forests have been effective in MENA and in FSU; these regions have been able to maintain all their forests in this scenario. The former has done so because it has sought to preserve its water resources and the latter because the extension of its productive lands has focused on areas where the permafrost has melted. In the OECD region, the reforestation phenomenon observed between 1990 and 2003 has increased up to 2050. It is therefore in the remaining three regions—ASIA, SSA and LAM—that the loss of forest surface areas has been concentrated. These regions have lost respectively 10%, 9% and 4% of the forest areas that they had in 2000. The annual rate of deforestation has increased slightly in ASIA, from 0.14% per annum between 1961 and 2000 to 0.22% per annum between 2000 and 2050. In contrast it has slowed down considerably in the other two regions (from 0.24 to 0.08% per annum in LAM and from 0.26 to 0.19% per annum in SSA).

Irrigated areas have been maintained in all the regions between 2000 and 2050 except in SSA where they have doubled (+7 million ha) and in ASIA where they have gained 6% (+9 million ha).

To recap, in the Agrimonde 1 scenario the sum of cultivated areas and pastures has increased by 2% in 2050 compared to 2000. As regards cultivated areas, in 2050 LAM and SSA have far from exploited their cultivable potential, and can afford to exploit only those lands with a high yield potential; their cultivated area is only 90% and 80% respectively of the land with a yield potential greater than 80% of the maximum attainable yield (VS). The OECD region cultivates 89% of the land with a yield potential greater than 60% of the maximum attainable yield

Table 6.2. Driving forces behind evolving land use in the Agrimonde 1 scenario

Region	Driving forces promoting growth of cultivated areas	Forces limiting the growth of cultivated areas
MENA		Saturation of the potential of cultivable land Accentuation of water stress due to climate change Necessity to preserve forest areas for hydrological functioning Artificialisation due to urbanisation
SSA	Reserves of cultivable land Limited capacity, in terms of governance, to curb the progression of the pioneering front on the savannah and Congo forest basin Rural and agricultural development policies: opening up of isolated areas and organisation of rural areas	Aridification due to climate change
LAM	Reserve of cultivable land Agricultural policy (research, training, development) Increasing production of agro-fuels	Aridification in Central America and the *Cerrado* due to climate change Growth of pressure groups to preserve the Amazon jungle
ASIA	Strong growth of agricultural population	Saturation of the potential of cultivable land Accentuation of water stress due to climate change in northern China Artificialisation due to urbanisation
FSU	Reserve of cultivable land Melting of the permafrost	Population decline
OECD	Reserve of cultivable land Melting of the permafrost in Canada Increasing production of agro-fuels	Reforestation Mechanisms to preserve protected species

(VS + S), and FSU 75 % of the land with a yield potential greater than 40 % of the maximum attainable yield (VS + S + MS). On the other hand, in this scenario ASIA and MENA are compelled to cultivate marginal land, representing respectively 34 and 4 million ha with a yield potential of less than 20 % of the maximum attainable yield (VS + S + MS + mS + NS).

Chapter 7
Food Crop Yields in 2050

Tévécia Ronzon

In the Agrimonde scenarios, each region's food resources are quantified on the basis of land use assumptions, food crop yields and the conversion of plant calories into animal calories. This chapter sets out the quantitative assumptions on food crop yields. First, the quantification principles opted for will be described, and then the yield assumptions will be presented for each region and, by aggregation, for the whole world on the 2050 timeline.

We refer to the data presented in the overview of the world food economy from 1961 to 2003 (Chap. 3).

The Principles for Quantifying Food Crop Yields

Our approach to building assumptions on yields—as in the case for our assumptions on cultivated areas—differs substantially from the approach in the Millennium Ecosystem Assessment (MA). Whereas in the Agrimonde scenarios the overriding idea is not *a priori* to balance biomass resources and uses, in MA yields are calculated using the IMPACT equilibrium model. In this model they depend on world commodity prices, labour and capital costs, as well as technological progress, itself determined by public and private research and development efforts, farmers' training, the development of infrastructures and markets, and capacity for irrigation (Box 7.1) (MA 2005b).

The Agrimonde GO scenario reproduces the yield gains of the MA Global Orchestration scenario. Note however that the only assumptions on yields available in the MA are those concerning cereals, distinguished by region. We therefore applied the same growth rate of yields to the variable "yields for all food crops combined" even though these assumptions become highly optimistic for all regions of the

T. Ronzon (✉)
IPTS, C/Inca Garcilaso, s/n 41092, Seville, Spain
e-mail: tevecia.ronzon@ec.europa.eu

S. Paillard et al. (eds.), *Agrimonde – Scenarios and Challenges for Feeding the World in 2050,*
DOI 10.1007/978-94-017-8745-1_7, © Éditions Quæ, 2014

Table 7.1 Food crop yields and annual growth rates: 1961–2000 and 2000–2050 in the Agrimonde scenarios[a]

Region	1961–2000			Agrimonde 1, low variant		2000–2050—Agrimonde Agrimonde 1, high variant		Agrimonde GO	
	Food crop yields[b] (kcal/ha/day)		Annual growth rate	Food crop yields[b] (kcal/ha/day)	Annual growth rate	Food crop yields[b] (kcal/ha/day)	Annual growth rate	Food crop yields[b] (kcal/ha/day)	Annual growth rate
	1961	2000	1961–2000	2050	2000–2050	2050	2000–2050	2050	2000–2050
World	8,607	18,703	2.01	20,027	0.14	30,462	0.98	32,940	1.14
MENA	4,921	12,836	2.49	14,500	0.24	17,970	0.68	21,362	1.02
SSA	5,027	9,460	1.63	11,750	0.44	18,920	1.40	23,133	1.80
LAM	9,041	18,688	1.88	23,500	0.46	37,376	1.40	36,494	1.35
ASIA	9,485	25,134	2.53	25,100	0	37,700	0.81	46,416	1.23
FSU	6,549	7,476	0.34	14,500	1.33	22,428	2.22	12,825	1.09
OECD	10,742	22,587	1.92	22,600	0	33,880	0.81	33,507	0.79

[a] The annual growth rates in the above graphs for Global Orchestration relate to the period 1997–2050 (MA 2005b). They differ slightly from those presented here for Agrimonde GO, which concern the period 2000–2050

[b] Between 1961 and 2003, cultivated land and food crop land areas were assumed to be the same since non-food crop areas were generally insignificant. Thus, yields are defined as production of plant food calories per ha of cultivated land in 1961 and 2000 while they are defined as production of plant food calories per ha of food crop land in 2050

world and, in this case, far higher than those chosen in the Agrimonde 1 scenario and presented in the current chapter (Table 7.1).

In Agrimonde 1 we chose to put forward a range of yields rather than making a single assumption with a set yield for each region. This made it possible to test the leeway of the system associated with yields, which is valuable in a foresight study intended to fuel reflection on long-term research orientations. Hence, if the low variant does not allow for a level of resources higher than or equal to the level of world use, it is possible to test the capacity to cover needs allowed for by the high variant and to draw conclusions on the size of the challenge facing research and innovation.

Box 7.1—Assumptions on cereal crop yields in the Millennium Ecosystem Assessment scenarios

Global Orchestration

This scenario is characterised by a sharp rise in yields between 2000 and 2050, both in developed and in developing countries, owing to major investments in agricultural research, a vast increase in irrigated areas, more efficient use of water and energy, and investments in support infrastructure. New techniques are mainly GMOs, more intensive crop farming, and increased use of fertilisers. Almost all farms, small and large alike, have become highly mechanised and industrial. Farmers who do not practise intensive farming—either by choice or on marginal lands—have very little weight in their country's economy and food production. Local knowledge is often replaced by standardised industrial methods.

TechnoGarden
Yield increases have been lower in rich countries, following the expansion of organic farming, than in poor countries. Investments in biotechnologies and other agricultural innovations have nevertheless been sufficient to generate significant yield gains. Likewise, improvements in animal production (average daily gain) have been weak in the rich countries, in particular because significant progress has already been made and because the demand for animal products has decreased. Innovations in animal reproduction have however enabled poor countries to substantially improve animals' average weight.

Order from Strength
In all countries, improvements in yields have been very weak from 2000 to 2050 since investments in infrastructure and in more efficient use of water and energy in agriculture have been insufficient. Yield gains have been obtained primarily through intensification and greater use of fertilisers. The lowest yield increases have been in OECD and FSU, while the highest have been in the developing countries. As customs duties in both rich and poor countries have pushed up the cost of agricultural technologies, poor farmers have not been able to acquire the techniques enabling them to maintain the fertility of land and other ecosystem functions. Without these techniques, their agricultural production has lagged and has consequently become more vulnerable to natural risks (diseases, pests, etc.). To maintain their income, farmers have enlarged their cultivated areas, especially on marginal land.

Adapting Mosaic
Yields increased at a moderate pace during the early years of the century and then slowed down in rich countries due to the large-scale adoption of organic farming. Yield increases have been greatest in developing countries owing to the successful implementation of climate change adaptation mechanisms. The poor countries have followed the movement, initiated in Europe and the USA, of environment-friendly farming and the marketing of healthy products. In 2030 organic produce and "produced naturally" goods account for 34% of the market in Europe and 21% in the USA. Environmental technologies, developed on the basis of local conditions and needs, have led to a gradual increase in the management of socio- ecological systems and natural resources. The diversification of skills and expertise has continued, owing to greater involvement by poor countries. Hence, the pace of technological progress aimed at using resources more efficiently and reasonably, and reducing pollution, has accelerated. (Source: MA 2005b)

To construct assumptions on yields, we proceeded in three stages.

1) The analysis of past trends and the identification of possible shifts. The analysis of regional productivity curves (shown Fig. 3.9), made it possible to some extent to assess each region's capacity to maintain, pursue or accelerate its yield gain rate. Two criteria in particular guided this analysis and allowed for the identification of possible changes in trends. The first was the form of the curve, which reflected the fact that the region under consideration was possibly experiencing a phase of technological progress if the curve was climbing, or stagnation if it was levelling out. The second was the yield level reached in 2000, which may suggest that the possibilities of yield gains had not yet been fully activated when the yield level in 2000 was particularly low.

2) Improvement of the consistency of the range of yields selected, with the assumptions of cultivated areas in Agrimonde 1, especially the yield potential of cultivated land of the different regions, and with the expected impacts of global warming on crops in each large ecosystem of those regions.

3) The analysis of regional yield assumptions and projections in existing agricultural foresight and forecast studies:

- Yield assumptions in MA scenarios up to 2050 (MA 2005b),
- Yield assumptions in the International Assessment of Agricultural Knowledge, Science and Technology for Development (IAASTD) up to 2050 (IAASTD 2009): high variant which corresponds to a large investment effort in agricultural knowledge, science and technology, and low variant which corresponds to a low effort,
- Yield assumptions in the Doubly Green Revolution (DGR) scenario of Michel Griffon up to 2050 (Griffon 2006),
- Baseline projections to 2020 by the International Food Policy Research Institute (IFPRI) (Rosegrant et al. 2001),
- Yield assumptions in the scenarios up to 2025 from the study "World Water and Food to 2025: Dealing with Scarcity" which results from a partnership between IFPRI and the International Water Management Institute (IWMI): "Business-As-Usual" (BAU), "Water Crisis" (CRI), "Sustainable Water Use" (SUS) (Rosegrant et al. 2002).

This analysis contributed to the formulation of assumptions, primarily by making it possible to identify yield increases associated in each exercise with models of agricultural production that respects the environment and preserves natural resources[1].

[1] Note that in his book Nourrir la Planète, Michel Griffon reasons on the basis of average cereal yield equivalents, and that the other exercises mobilised present only cereal yield trends. However, in the Agrimonde scenarios we discussed yields for all types of food crops combined, expressed in kcal/ha/day.

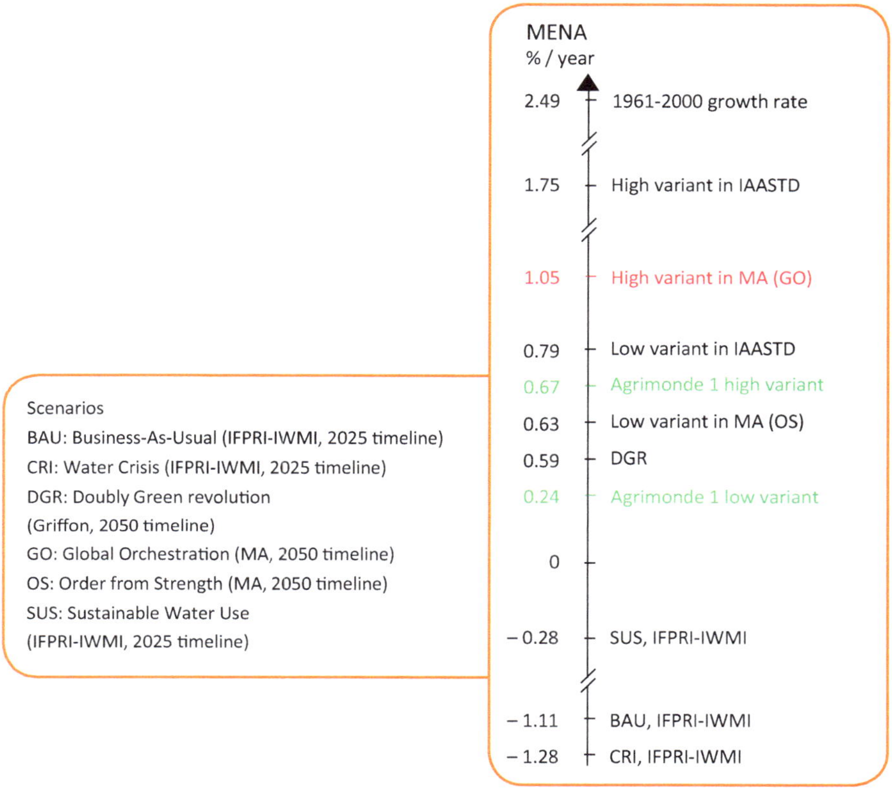

Fig. 7.1 MENA: References for the quantification of yields in 2050

Food Crop Yields in Middle East—North Africa in 2050

Between 1961 and 2000, according to Agribiom estimates based on the United Nations Food and Agriculture Organization (FAO) data, food crop yields in Middle East—North Africa (MENA) grew considerably and are still on the rise (+2.49% per annum for all crops combined, i.e. a 160% increase over 40 years) (Fig. A2.6). The growth curve for yields in this region does not seem to level off. It therefore seems that this region still has a considerable potential for increasing its yields. The analysis of these past trends raises the following question: up until 2050, will the MENA region be able to maintain the same rate of crop yield gains as between 1961 and 2000?

Existing scenarios all foresee a deceleration of crop yield gains in this region in the future, but to varying degrees (Fig. 7.1). While Michel Griffon's Doubly Green Revolution (DGR) scenario (Griffon 2006) and the MA scenarios (MA 2005b), as well as the IFPRI baseline projection for 2020 (Rosegrant et al. 2001) foresee gains of less than or close to 1% per annum (i.e. +64% over 50 years), the IAASTD high variant is far more optimistic (a +1.75% gain per annum, i.e. a 160% increase of yields between 2000 and 2050). The IAASTD ascribes these gains to massive investments in agricultural research, development and training (IAASTD 2009). Only the IFPRI-IWMI scenarios for 2025, in the study "World Water and Food to 2025: Dealing with Scarcity", foresee a drop in current yield levels, moderate in the case of the SUS scenario (−0.28% per annum between 1995 and 2025), to high in the BAU and CRI scenarios (respectively −1.28% and −1.11% per annum between 1995 and 2025) (Fig. 7.1) (Rosegrant et al. 2002). Studies that focus on climate change anticipate a decrease in cereal yields due to the aridification expected in this region (see the results of the studies of Parry et al. (2004) and Cline (2007) in Box 7.2).

Box 7.2—Expected impacts of climate change in Middle East—North Africa

Results of the IPCC's work show fairly homogeneous impacts of climate change in MENA until the end of the century (IPCC 2007c):

- Mean annual temperatures will have risen by 3 to 5 °C in 2080–2099 compared to 1980–1999, except in the Mediterranean Basin where they are expected to rise by only 3 to 3.5 °C. These temperature increases will be felt most intensely in summer when they will be between 4 and 5 °C higher throughout the Western Sahara and Middle East.
- The Mediterranean Basin will tend to become drier, with a 30% decrease in mean annual rainfall and a 50% decrease in summer rainfall.
- Only the south of Saudi Arabia, Yemen and Oman will experience increases in rainfall, with a mean annual increase of up to 20%.

Parry et al. (2004) calculate yield losses in the southern Mediterranean countries of 0 to 2.5% in 2050, compared to 1990, in the IPCC scenario with the lowest CO_2 emissions (B1) (i.e. 0 to −0.04% per annum) and of 2.5 to 5% (i.e. −0.04 to −0.08% per annum) in the IPCC scenario with the highest CO_2 emissions (A1FI). The yield losses that they calculate will be greater in the eastern Mediterranean countries: −0.04% to −0.08% per annum in B1, and −0.08% to −0.17% per annum in A1FI. The study is more optimistic for Turkey, which will experience yield gains (0 to +0.04% per annum in B1 and +0.04 to +0.08% per annum in A1FI).

Cline (2007) estimates production losses at −0.26% per annum by 2080 without carbon fertilisation, reducible to −0.10% per annum with carbon fertilisation.

In the Agrimonde 1 scenario, water stress, which is assumed to have intensified with global warming, has been a factor leading to stagnation of cultivated and irrigated surface areas. Only 7 million additional hectares have been planted between 2000 and 2050, and this has been on land with a yield potential of less than 20 % of the maximum attainable yield[2]. Water stress has also been a factor limiting land productivity. Thus, the assumption in this scenario is growth in productivity of food crops (all types combined) of between 0.24 and 0.67 % per annum in MENA (i.e. an increase from 12,800 kcal/ha/day in 2000 to a range of 14,500 to 18,000 kcal/ha/day in 2050). The low variant is more cautious than the assumptions made by Michel Griffon (DGR) and the MA but less alarmist than those of Parry et al. (2004) based on the expected effects of climate change in this region (Box 7.2). It seems more cautious to credit the low variant of the Agrimonde 1 scenario which assumes that climate change adaptation strategies, especially water stress, will have been implemented by 2050 (Chap. 9).

Food Crop Yields in Sub-Saharan Africa in 2050

According to Agribiom estimates based on FAO data, in sub-Saharan Africa (SSA), land productivity in food plant calories almost doubled between 1961 and 2000 (5,000 plant kcal/ha/day in 1961 to 9,500 in 2000, i.e. an annual growth rate of 1.63 %). Notwithstanding a slight downturn since 2000, the growth curve of yields in this region does not seem to level off (Fig. A2.6). Moreover, in 2000 the crop yields in SSA are amongst the lowest in the world. Yields in Asia, the highest in the world, are five times greater. SSA still has a high yield growth potential, but will this region be able to double its crop yields again by 2050?

Most of the scenarios that we analysed foresee annual growth rates that would enable SSA to double cereal yields by 2050 (MA Adapting Mosaic and TechnoGarden scenarios, as well as the IFPRI-IWMI BAU scenario for 2025 and the IFPRI reference scenario for 2020), or to multiply them by 2.5 (MA Global Orchestration scenario) or even to triple them (high variant of the IAASTD). Only Michel Griffon's DGR scenario and the most pessimistic scenarios in terms of economic growth (the MA Order from Strength scenario), water availability (the IFPRI-IWMI's CRI scenario for 2025) or progress in research and its transfer (low variant of the IAASTD) do not foresee such large yield increases[3] (Fig. 7.2).

For SSA, Michel Griffon proposes a scenario in which increases in regional agricultural production between 2000 and 2050 are a result of the extension of agricultural surfaces rather than yield gains. Consequently, in his DGR scenario, yields have increased only 0.62 % per annum (almost 40 % in 50 years), essentially owing to improved productivity of rainfed crops (+40 % of rainfed crop yields by 2050, compared only +15 % for irrigated crops). He highlights the fact that other factors have to be implemented for development of the SSA region: improved governance, development of social capital, and infrastructure for market access.

---

[2] According to the typology of the GAEZ study (Fischer et al. 2000, 2001, 2002).

[3] In MA and Agrimonde, the SSA region is larger than in IAASTD. It also encompasses Djibouti, Mauritania, Mayotte, Somalia, Sudan, Reunion Island and Sainte Hélène.

Fig. 7.2 SSA: References for the quantification of yields in 2050

In the Agrimonde 1 scenario, cultivated areas in SSA have grown by 150 million ha, on land with a yield potential greater than 60 % of the maximum attainable yield (see note 73). Irrigated areas have doubled and have partly contributed to the growth of crop yields. SSA has a high yield gain potential, and even if yields were to double again (all food crops combined), as they did between 1961 and 2000, this region would still have a lower plant calorie productivity than that of Latin America (LAM) in 2000. However, studies on climate change are not optimistic for this region of the world (Box 7.3), where certain ecosystems are already experiencing ecological crises (the highlands, in particular) and the yield curve has been levelling off since the end of the twentieth century. These criteria point to a slowdown in yield gains. Therefore, the food crop yield assumptions in the Agrimonde 1 scenario for SSA range from a gain of 25 % (0.44 % per annum, roughly the same as that of the DGR scenario) to a gain of 100 % in 2050 (that is, 1.40 % per annum). This amounts to a yield gain of 9,500 kcal/ha/day in 2000 to between 11,750 and 18,900 kcal/ha/day in 2050. The range of yields chosen assumes that certain changes and adjustments will have been made in SSA by 2050. These are discussed in Chap. 9.

Box 7.3—Expected climate change in sub-Saharan Africa
The work of the Intergovernmental Panel on Climate Change (IPCC) shows that in SSA climate change in the twenty-first century will have different effects in the various zones (IPCC 2007c):

- Southern Africa will tend to become more arid, with mean annual temperature increases of up to 4 °C in 2080–2099 compared to 1980–1999, and a mean annual loss of precipitation of 5 to 10 % over the same period. The months of June-July-August will experience a decrease in precipitation of up to −50 %.
- On the east coast and around the Gulf of Guinea, mean annual temperature increases will be limited to between 2 and 3 °C in 2080–2099 compared to 1980–1999, and will be accompanied by increases in precipitation. The mean annual rainfall will increase by 0 to 5 % over the same period in the Gulf of Guinea and by 5 to 20 % from eastern Tanzania to Ethiopia, with peaks during the first quarter of up to +50 % in the Sudan-Ethiopia area.

Globally, climate change could lead to a shortening of the vegetative period (Van de Steeg 2009), especially in Central and Eastern Africa. More intense rainfall could exacerbate problems of erosion (ibid.) and the main food crop yields could decline across the continent (Challinor et al. 2007, Lobell et al. 2008). Arid and semi-arid areas could increase by 5 to 8 % by 2080. The situation is expected to be most critical in the more populous areas of central, west and southern Africa (IPCC 2007b). On the other hand, climatic conditions could allow for a lengthening of the vegetative period on the Ethiopian highlands (Thornton et al. 2005).

Parry et al. (2004) calculate a yield loss of 2.5 % in 2050 in the region as a whole, compared to 1990 in the IPCC scenario with the lowest CO_2 emissions (B1), which could be as high as 5 % in southern Africa, Ivory Coast, Ghana and Cameroon, and even 10 % in Nigeria. In the case of the scenario with the highest CO_2 emissions (A1FI), losses are expected to be between 2.5 and 5 % (i.e. between −0.04 and −0.08 % per annum) in the region as a whole, and between 5 and 10 % (i.e. −0.08 % and 0.17 % per annum) in Nigeria and Guinea Bissau.

Cline (2007) calculates production losses of −0.36 % per annum by 2080 without carbon fertilisation, reducible to −0.21 % per annum with carbon fertilisation.

Food Crop Yields in Latin America in 2050

Between 1961 and 2000 food crop yields doubled in Latin America (LAM) according to Agribiom estimates based on FAO data (9,000 plant kcal/ha/day in 1961 to 18,700 in 2000, i.e. a growth rate of 1.88 % per annum). Far from reaching a ceiling, the increase in yields accelerated between 1992 and 2003, to the extent that yields almost attained 50 % growth during this short period. In 2003 food crop yields in this

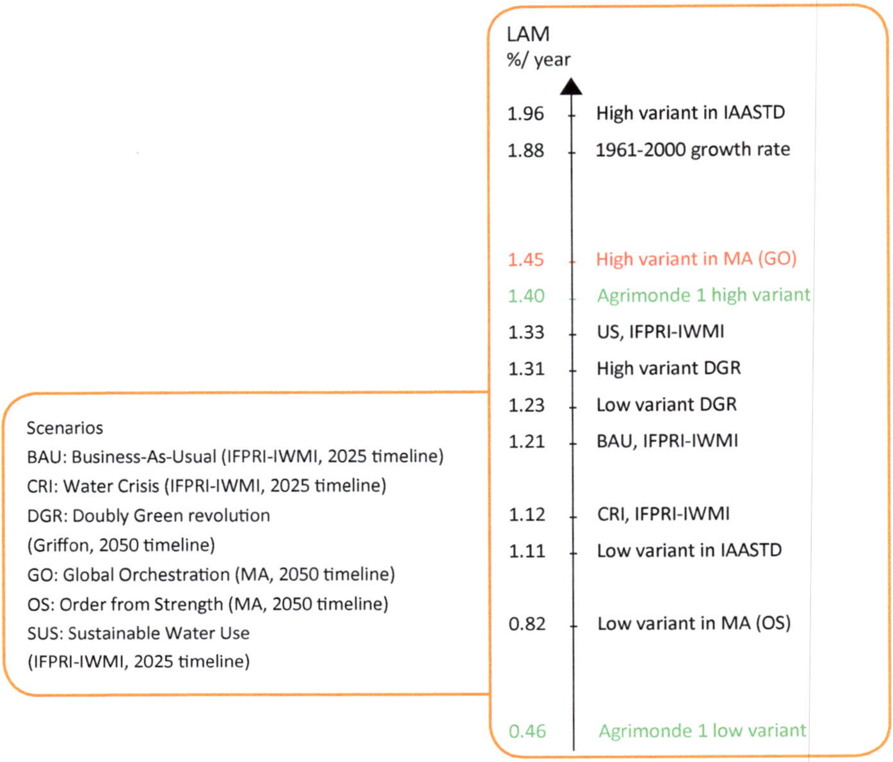

Fig. 7.3 LAM: References for the quantification of yields in 2050. (Michel Griffon proposes a high yield variant and a low yield variant for LAM but only one variant for the other regions (Griffon 2006).)

region were amongst the highest in the world and even greater than those of OECD (23,000 kcal/ha/day compared to 21,900 kcal/ha/day in OECD in 2003) (Fig. A2.6). The analysis of these past trends raises the following question: does LAM have the capacity to double its food crop yields again between 2000 and 2050?

LAM's strong potential for cereal yield gains seems to be taken for granted in existing foresight scenarios. Apart from crisis scenarios (the MA Order from Strength scenario, the low variant of the IAASTD, and the CRI scenario of the IFPRI-IWMI for 2025), all assume annual growth rates that would almost double yields over the next 50 years (from + 1.21 % per annum in the IFPRI-IWMI BAU scenario for 2025, to + 1.50 % per annum in the IFPRI reference scenario for 2020. The IAASTD high variant goes even further: with a yield growth rate of 1.96 % per annum, it represents a 164 % increase over 50 years (Fig. 7.3).

In Michel Griffon's DGR scenario, many justifications and conditions accompany the assumption of doubled yields. "For viable family farms, as for average and large farms in humid tropical areas and grasslands, the investments that have to be made are low in relation to the expected result. It is therefore realistic to believe that the productive potential could be expressed without agricultural support policies being necessary. On the other hand, the sector of very small farms and landless rural workers can use ecological intensification techniques only if programmes enable them to have access to land, to expand their farms, and to have access to credit and to the market" (Griffon 2006).

In the Agrimonde 1 scenario, cultivated surfaces in LAM have gained 150 million ha on land with a high agricultural potential, i.e. with a yield potential greater than 80 % of the maximum attainable yield (Chap. 6). Irrigated areas remain stable and cannot explain yield gains in this region. The analysis of past trends argues for a continued increase of food crop yields for the period 2000–2050 in LAM. Moreover, the specific characteristics of this region, i.e. the large amount of available space, low demographic density, available technologies, and relative political stability, support this analysis. The doubling of yields does not seem impossible. However, the state of the soil is already a problem, especially in the *Cerrado* and on the forest pioneering fronts where extensive crop and livestock farming follow deforestation without any concern for regenerating the soil's fertility. As global warming is likely to make ecosystems in Central America, in the north-east of Brazil and on the *Pampas* even more fragile (Box 7.4), a lower variant of yield gains can be considered. Consequently, the Agrimonde 1 scenario for LAM assumes food crop yield increases in 2050 will range from just under 30 % (0.46 % per annum) to 100 % (or 1.40 % per annum, i.e. consistent with the projections in most of the scenarios referred to). This amounts to a jump from a yield level of 18,700 kcal/ha/day in 2000, to a level situated between 23,500 and 37,400 kcal/ha/day in 2050. The range of yields selected in the Agrimonde 1 scenario would require a number of changes and adaptations in LAM by 2050. These are discussed in Chap. 9.

Box 7.4—Expected climate change in Latin America

For 2100, the IPCC (IPCC 2007c) foresees a sharp rise in temperatures in LAM:

- in Central America, accompanied by reduced precipitation, especially in summer (between +2.5 and +4 °C, and −5 to −15 % of the mean annual precipitation in 2080–2099 compared to 1980–1999, and up to −30 % in summer),
- in the Amazon Basin, the temperature rise will be coupled with a very sharp increase in precipitation (up to +4 °C and between 0 and +5 % of the mean annual precipitation, and up to +5 °C in December-January-February).

Global warming will be more moderate:

- in the north-east of Brazil, which will however receive considerably less rain (+2.5 to 3 °C and −5 to −50 % of the mean annual rainfall),
- in the south of the continent (Chile and Argentina), where rainfall will decrease in Patagonia, especially in December-January-February (+1.5 to 3 °C and 0 to −20 % of the mean annual rainfall),

- in the areas of the Pampas which will receive more rainfall (+2 to 3 °C and +5 to +15% of the annual mean, and up to +20% in December-January-February),
- in the Andes which will also receive more rain (a mean annual increase of +2.5 to 3 °C and +5 to 15%, and up to +20% in December-January-February).

By 2050, desertification and salinisation could affect 50% of the land of the LAM region (FAO 2004). The stress caused by intensified heat, associated with drier soil, could reduce yields by a third in the tropical and sub-tropical zones where crops are already reaching their maximum point of tolerance (FAO 2001). Coffee production areas and yields could also decrease due to climate change. Moreover, despite a wide variability in projected yields, all studies agree on a decrease in rice yields after 2010 and an increase in soy yields when the effects of CO_2 are integrated.

The study by Parry et al. (2004) shows that the impact of climate change on yields in 2050 in LAM is likely to be relatively evenly spread out. Whether in the IPCC scenario with the lowest CO_2 emissions (B1) or in that of the highest CO_2 emissions (A1FI), yield losses are evaluated within a range of 2.5 to 5% in 2050 compared to the level in 1990 for the entire continent (i.e. between −0.04 and −0.08% per annum), with some countries more affected (between 5 and 10%, or between −0.08% and −0.17% per annum in Ecuador, Guyana, Surinam and Uruguay). Only Argentina is expected to have increasing yields, from 0 to 2.5% in scenario B1 and from 5 to 10% in scenario A1FI.

Cline (2007) estimates production losses at −0.30% per annum by 2080 without carbon fertilisation, reducible to −0.15% per annum with carbon fertilisation.

Food Crop Yields in Asia in 2050

According to Agribiom estimates based on FAO data, between 1961 and 2000, food crop yields were multiplied by 2.6 in Asia (ASIA) (from 9,500 kcal/ha/day to 25,100). Today they are the highest in the world in the Agrimonde geographical zoning. The plant calorie per hectare productivity curve nevertheless seems to have stagnated in the last 10 years (Fig. A2.6). Can food crop yields still increase by 2.53% per annum during the period 2000–2050, as in 1961–2000? Or is the stagnation that began in the early nineties set to continue?

None of the foresight scenarios to which we referred maintains the yield gains obtained by the green revolution in ASIA (Fig. 7.4). At best, the IWMI-IFPRI's SUS scenario assumes a growth rate of 1.32% per annum up to 2025. If this rate were to continue until 2050, it would mean that cereal yields doubled in 50 years but also at a rate only half that of the period 1961–2000 (all crops combined). The high

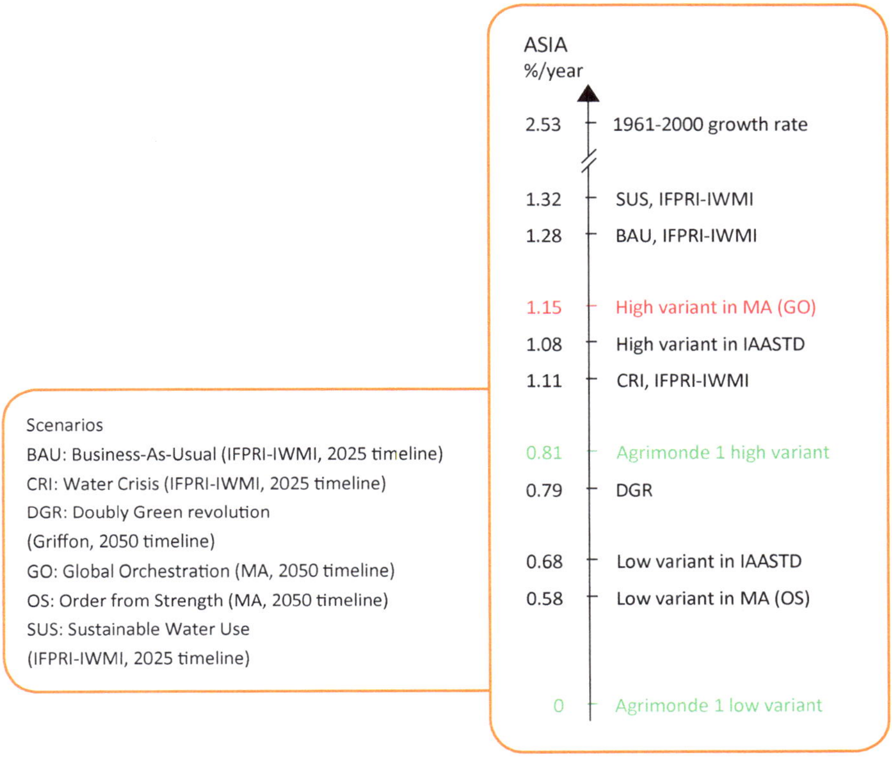

Fig. 7.4 ASIA: References for the quantification of yields in 2050

variants of the MA (Global Orchestration) and IAASTD scenarios are more moderate and project a cereal yield increase of around 75 % in 50 years.

Yet no scenario foresees that the stable yield levels observed over the past 10 years will remain unchanged. Even the most pessimistic (the MA Order from Strength and the IASSTD low variant) are based on a yield increase of 35 to 40 % by 2050 (respectively 0.58 and 0.68 % per annum).

Michel Griffon's assumption of a doubly green revolution is intermediate: +50 % by 2050, i.e. 0.79 % per annum. He notes that the productive effort "should be achieved by a large number of small family farms of which most are poor households". According to him, "the green revolution agricultural policy included most of these farms in a market economy, but the lack of capital made them very financially vulnerable. This economy is therefore highly dependent on prices and subsidies for fertilisers and pesticides. Hence, a productivity increase cannot be envisaged without profound stimulation by public policies" (Griffon 2006).

In the Agrimonde 1 scenario, cultivated areas in ASIA have grown by about 100 million ha between 2000 and 2050. This has been made possible by exploiting marginal lands: some 70 million ha have been cultivated on land with a yield potential of less than 40% of the maximum attainable yield, and the rest on land with a yield potential of less than 20% of the maximum attainable yield (see note 73). Irrigation has grown too slowly (some 10 million ha between 2000 and 2050) to fully explain these gains. It can be foreseen moreover that the delta regions with irrigated rice fields—South Vietnam, Bangladesh, Yellow River—will experience a new situation with the rise in sea levels which could lead to their salinisation. Crops in these deltas could also be threatened by the increasing frequency of violent climatic events such as cyclones (Box 7.5), and by the reduced capacity of watersheds upstream to retain water (reduction of plant coverage and forests in mountainous areas). The green revolution seems moreover to have attained its limits in these regions. In contrast, in ASIA there are still areas of rainfed crops where there is room for yield increases, as for instance in the states of Bihar or Jharkhand in the Union of India. Hence, the food crop yield assumption in the Agrimonde 1 scenario for ASIA corresponds to a range of 0 to +50% increases (i.e. 0.81% per annum, as in the DGR scenario of Michel Griffon). This is equivalent to a yield increase of 25,100 kcal/ha/day in 2000 to between 25,100 and 37,700 kcal/ha/day in 2050. However, given that over the past 10 years productivity has scarcely increased, we cannot assume that yields will increase, even by 50%, without major innovations. These are discussed in Chap. 9.

Box 7.5—Expected climate change in Asia

Temperature increases should be far higher than the mean global increase in northern China, Mongolia and on the Tibetan Plateau (an annual mean of +3.5 to +5 °C in 2080–2099 compared to 1980–1999, with peaks of +7 °C in the months of December-January-February on the border with the former Soviet Union). They will be attended by an increase in precipitation especially in winter (an annual mean of between +5 and +15%, and up to +50% in December-January-February) (IPCC 2007c).

The rise in temperatures is expected to be more moderate in the rest of the region (an annual mean of +2 to +3.5 °C) where rainfall is expected to increase by an annual mean of 0 to 10%. The Indochinese peninsula may receive less rainfall in winter (up to −20% in Myanmar in December-January-February) (IPCC 2007c). Heat waves in summer will be longer, more frequent and more intense, and the cold periods will be shorter (IPCC 2007a).

Extreme rains and tropical cyclones are expected to become more frequent, while the monsoon will weaken.

Many studies have suggested that global warming could entail substantial reductions of cereal production before the end of the century. Yields could increase by 20% in East and South-East Asia, but decrease by 30% in South Asia (IPCC 2007b). In particular, rice production could decrease by 3.8% by 2100 under the combined effects of heat and water stress (Murdiyarso 2000).

The study by Parry et al. (2004) shows that Asian cereal yields are not likely to be strongly affected by climate change. Overall, their figures range from 0 to +2.5% between 1990 and 2050 (i.e. between 0 and +0.04% per annum) in the IPCC scenario with the highest levels of CO_2 emissions (A1FI) and by 0 to −2.5% in the IPCC scenario with the lowest emissions (B1) (i.e. between 0 and −0.04% per annum). Yield losses are expected to be amplified in India (between −2.5 and −5% in the two scenarios, or −0.04 and −0.08% per annum), while the Indochinese peninsula will experience gains of +2.5 to 5% (between +0.04 and +0.08% per annum).

Cline (2007) calculates production losses at −0.23% per annum by 2080 without carbon fertilisation, reducible to −0.08% per annum with carbon fertilisation.

Food Crop Yields in the Former Soviet Union in 2050

In the former Soviet Union (FSU), the plant calories per hectare productivity curve increased overall from 1961 to 1990, and then dropped after the opening of the Eastern Bloc. In 2000 it had still not restored its 1990 level (11,000 kcal/ha/day in 1990, compared to 7,500 kcal/ha/day in 2000) (Fig. A2.6). Historical data series are however highly chaotic and their reliability is doubtful.

Michel Griffon assumes in his DGR scenario that there will be a 30% increase in cereal yields on the plains of the CIS (the Commonwealth of Independent States) between 2000 and 2050. The MA scenarios envisage an increase of cereal yields of between 20% and 45% over the same period. The IFPRI baseline scenario for 2020 also assumes a growth rate within this range. Comparison with the IAASTD projections is difficult because of their geographical zoning, in which FSU belongs to two regions: North America-Europe (NAE) and Asia. In these two regions the high and low variants of the growth rates of cereal yields are respectively +0.71% and +1.33% per annum in NAE and +0.68 and +1.08% per annum in Asia. The IFPRI-IWMI scenarios for 2025 include FSU in a much broader category of "Developed Countries". The projected growth rates are between +0.58% per annum for the CRI scenario and +0.88% per annum in the SUS scenario (Fig. 7.5).

In the Agrimonde 1 scenario, cultivated areas have increased by about 100 million ha between 2000 and 2050, partly on former pastures which have moved north to areas where the permafrost has melted. It is probable that these lands have a very moderate agricultural potential (Box 7.6). Irrigated areas have remained stable and therefore cannot fully explain the increasing yields. Initially we considered it realistic to assume that in 2050 the level of food crop yields in FSU would attain the OECD level of 2000– which would imply a threefold yield increase. This assumption, representing an annual growth rate of 2.26%, is however situated far higher than those in the various scenarios referred to above. The food crop yield assumption in the Agrimonde 1 scenario for FSU thus corresponds to a range, from a 100% (+1.33% per annum) to a 200% (+2.22% per annum) yield increase between 2000 and 2050. This amounts to an overall yield increase of 7,500 kcal/ha/day in 2000 to a level between 14,500 and 22,400 kcal/ha/day in 2050. The conditions required for

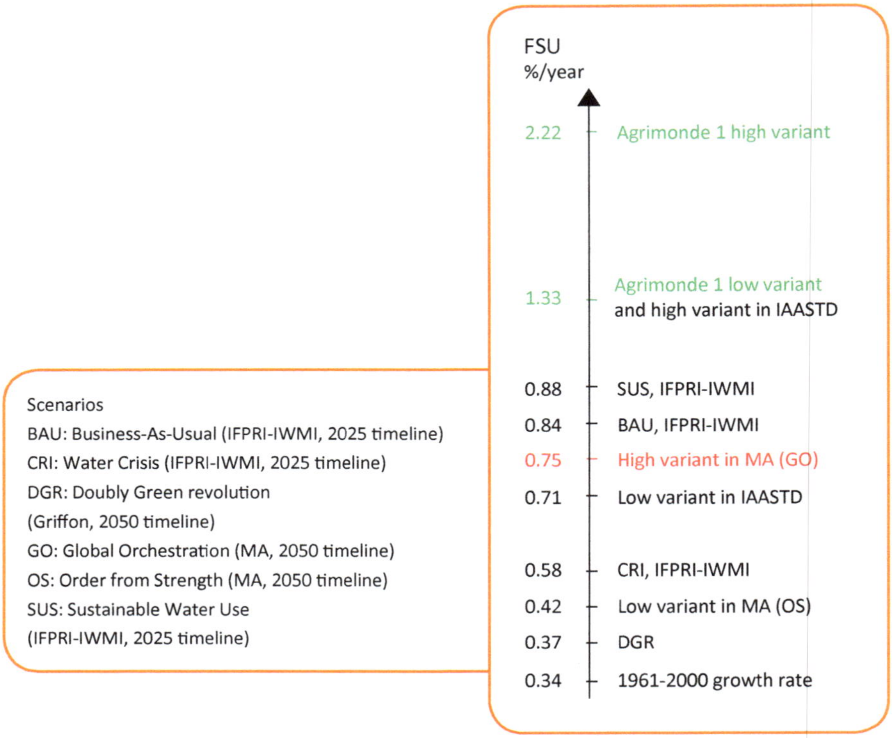

Fig. 7.5 FSU: References for the quantification of yields in 2050

Box 7.6—Expected climate change in the former Soviet Union

Temperature increases in FSU are expected to be far greater than the global mean, in all the IPCC scenarios.

They will be particularly high in the regions situated north of the 60th parallel (an annual mean of +4 to +5 °C in the western part in 2080–2099 compared to 1980–1999, and of +4 to +7 °C in the east). These warmer temperatures will be combined with a mean annual increase in rainfall and snow, of between 15 % and 30 %, according to a west-east gradient. These trends will peak in winter, when temperature increases will be as high as +10 °C and precipitation increases will attain 50 % in the months of December-January-February (IPCC 2007c).

Between the 50th and 60th parallels north, temperature increases will be lower: between 3.5 and 4 °C in the western part of FSU, and between 4 and 5 °C in the eastern part. They will also be coupled with mean annual rainfall increases of up to 15 % in 2080–2099 compared to 1980–1999, or even +20 % in the winter (months of December-January-February) (IPCC 2007c).

Finally, south of the 50th parallel, in central Asia, temperatures will experience a mean annual increase of 3.5 to 4 °C in 2080–2099 compared to 1980–1999. Precipitation will also increase north of the Caspian Sea (up to + 10 % and + 15 % in winter) but will decrease over the Caspian Sea (up to − 10 % and even − 30 % in summer) (IPCC 2007c).

The study by Parry et al. (2004) shows that cereal yields in FSU will on the whole be affected by climate change. They will decrease by 5 to 10 % between 1990 and 2050 (i.e. between − 0.08 and − 0.17 % per annum) in both of the IPCC scenarios: A1FI (with the highest CO_2 emissions and B1 with the lowest CO_2 emissions).

this assumption to come to fruition, especially in terms of manpower needs, will be discussed in Chap. 9.

Food Crop Yields in the OECD-1990 Region in 2050

According to Agribiom estimates based on FAO data, in the OECD-1990 region (OECD), the productivity of the land in terms of plant calories doubled between 1961 and 2000, from 10,700 plant kcal/ha/day in 1961 to 22,600 in 2000 (i.e. an annual growth rate of 1.92 % between 1961 and 2000). In 2000, the food crop yields in the region were among the highest in the world. Their growth nevertheless seems to have

Box 7.7—Expected climate change in OECD-1990
Temperature increases in North America will be particularly high north of the 50th parallel north, where mean annual increases of 3.5 to 7 °C in 2080–2099 as compared to 1980–1999 will be recorded, coupled with an increase in precipitation of 1 to 30 %. This temperature rise will be more pronounced in winter when it could attain + 10 °C, whereas in summer it will remain in the range of 3 to 3.5 °C (IPCC 2007c). South of the 50th parallel north, overall temperature increases will be +3.5 to +4 °C with a north-south gradient, but will remain between +3 and +3.5 °C on the coast. Precipitation could increase up to 15 % except in the south of the US where it could decrease by 30 %. In summer the entire area will receive less rain (up to −30 % in the months of June-July-August) (IPCC 2007c).

The study by Parry et al. (2004) shows that cereal yields in the US could drop by 0 to 2.5 % (i.e. between 0 and − 0.04 % per annum) and those in Canada could increase by 5 to 10 % (i.e. between 0.08 and +0.17 % per annum) in the period from 1990 to 2050 in both of the IPCC scenarios: A1FI (with the highest level of CO_2 emissions) and B1 (with the lowest level of CO_2 emissions).

In Europe, on the basis of mean annual temperatures, global warming will be felt increasingly as one moves inland. Whereas the Atlantic coast is likely to experience a limited increase in temperatures of between 2.5 and 3 °C, and the Mediterranean coast of between 3 and 3.5 °C, the more continental countries of eastern and northern Europe will experience temperature increases of between 3.5 and 7 °C. In summer these increases will however be concentrated around the Mediterranean Basin (between 3.5 and 4 °C, whereas the rest of Europe is not expected to experience temperature rises of more than 3.5 °C) (IPCC 2007c). Changes in precipitation will follow a pattern with northern Europe on the one hand and southern Europe on the other, with the dividing line corresponding to the 50th parallel north. North of this parallel, mean annual rainfall will increase according to a north-south gradient, by between 5 and 20 %. South of this parallel, it will decrease by 0 to 30 %. The dividing line will move up to the 55th parallel north in summer and to the 45th parallel north in winter (IPCC 2007b).

The study by Parry et al. (2004) shows that cereal yields in eastern Europe could be affected by climate change and decline by 5 to 10 % between 1990 and 2050 in the IPCC scenario with the highest CO_2 emissions (A1FI) and by between 10 and 30 % in the IPCC scenario with the lowest emissions (B1) (i.e. between −0.08 and −0.17 % per annum in A1FI and between −0.17 and −0.59 % per annum in B1). On the other hand, they will increase in western Europe by 5 to 10 % in A1FI and by 2.5 to 5 % in B1 (i.e. between +0.08 and +0.17 % per annum in A1FI and between +0.04 and −0.08 % per annum in B1).

In Australia, global warming will be felt intensely in the centre of the country (a mean annual increase of 3 to 4 °C) compared to the coast (between 2 and 2.5 °C on the Indian Ocean coast and between 2.5 and 3.5 °C on the other coasts). Whereas there will be a mean annual increase in rainfall of 0 to +5 % in the north-eastern part of the country, in the south-west rainfall is expected to decrease by 5 to 20 % (IPCC 2007c). In New Zealand, temperatures will increase relatively little (an annual mean of between +1.5 and +2.5 °C) and rainfall could increase up to 10 % in the north (IPCC 2007c). The study by Parry et al. shows that cereal yields in Australia and New Zealand will increase by 2.5 to 5 % between 1990 and 2050 (i.e. between +0.04 and +0.08 % per annum) in the IPCC scenario with the lowest CO_2 emissions (B1) and 5 to 10 % (i.e. between +0.08 and +0.17 % per annum) in the IPCC scenario with the highest emissions of CO_2 (A1FI).

Cline (2007) calculates production losses in developed countries at −0.26 % per annum by 2080 without carbon fertilisation, reducible to −0.10 % per annum with carbon fertilisation.

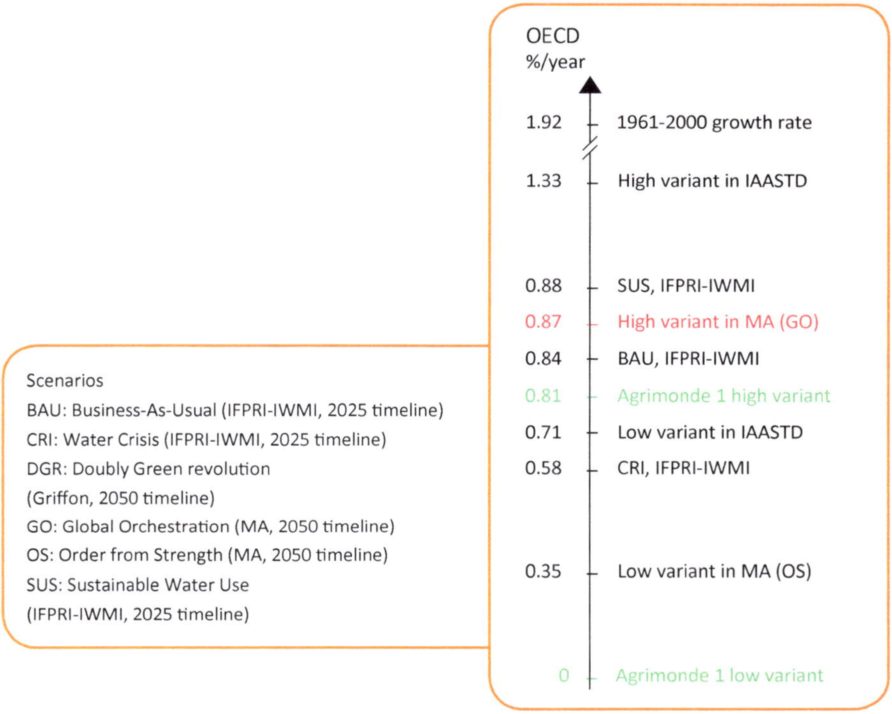

Fig. 7.6 OECD: References for the quantification of yields in 2050

slowed down in recent years (Fig. A2.6). Can food crop yields still increase in the period 2000–2050 at the rate observed between 1961 and 2000? Or will the deceleration initiated in recent years continue? What role will climate change play (Box 7.7)? Of the scenarios referred to above (Fig. 7.6), only the IAASTD high variant still foresees a doubling of cereal yields by 2050 (i.e. +1.33 % per annum over 50 years). This scenario projects increasing investments in agronomic research, development and training. The other scenarios appear to converge more towards a new yield gain of 50 % (between 0.84 and 0.88 % per annum over 50 years), or even of 40 % in the IFPRI baseline scenario for 2020 and the low variant of the IAASTD scenario (+0.71 % per annum). Finally, the crisis scenarios envisage a more drastic deceleration of yield growth in OECD, with yield gain assumptions of +0.35 % per annum up to 2050 (i.e. +20 %) in the MA Order from Strength scenario and +0.58 % per annum up to 2025 in the IFPRI-IWMI CRI scenario (i.e. the equivalent of +35 % over 50 years)[4].

In the Agrimonde 1 scenario, cultivated areas have expanded by close to 80 million ha on land with a yield potential greater than 60 % of the maximum attainable yield (see note 73). Irrigated areas have remained stable and cannot therefore be

[4] Yield gains in OECD are not specified in Michel Griffon's scenario. The scenarios for 2050 of the IAASTD, the IFPRI-IWMI for 2025 and the IFPRI for 2020 make no assumptions for the OECD region as such. The IAASTD includes it in the North America—Europe region, while the IFPRI-IWMI 2025 and the IFPRI 2020 include it in "developed countries".

considered as a factor of yield gains. Reserves for crop yield increases (all crops combined) in OECD do exist. In 2000 the countries of Eastern Europe had not yet attained their 1990 level, and some countries like Australia and New Zealand are characterised by production systems that are still highly extensive. Finally, agronomic research can allow for further yield gains in varieties grown in this region, by way of either plant selection or improvements in production practices. It therefore does not seem unrealistic to envisage a further 50 % increase in crop yields over 50 years, as in most of the scenarios referred to. In a sustainable development scenario the objective of quality products from an organoleptic, sanitary and environmental point of view could however result in less intensification of production systems. The assumption of food crop yields in the Agrimonde 1 scenario for the OECD region therefore ranges from stable yields to a + 50 % increase (i.e. 0.81 % per annum, as in the MA TechnoGarden and Global Orchestration scenarios, and in the IFPRI-IWMI BAU and SUS scenarios for 2025). This represents a steady level of production of 22,600 kcal/ha/day in the period from 2000 to 2050, for the low variant, and an increase to 33,900 kcal/ha/day in the same period, for the high variant. The conditions in which this assumption would come to fruition are discussed in Chap. 9.

Food Crop Yields in the World in 2050

According to Agribiom estimates based on FAO data, on a global scale, the production of plant calories per hectare and per day doubled between 1961 and 2000, from 8,600 kcal in 1961 to 18,700 kcal in 2000 (Fig. A2.6). Behind this mean global curve, each region presented its own dynamics, and regional trends reflect the differing expressions of multiple factors ranging from technological progress (plant variety improvements, intensification of inputs, expansion of irrigated crops) to the organisation of the agricultural sector (agricultural policies, access to information, to credit, to land, etc.)—and of course the initial yield level.

Despite the diversity of trajectories in the various regions of the world, two main groups seem to emerge in 2000 (Fig. 3.9):

- a group of three regions that consistently have the lowest levels of food crop yields throughout the period: SSA, MENA, and FSU (from 4,900 to 6,500 kcal/ha/day in 1961 and from 7,500 to 12,800 kcal/ha/day in 2000),
- a group of regions with the highest food crop yields throughout the period: LAM, ASIA and OECD (between 9,000 and 10,700 kcal/ha/day in 1961 and from 18,700 to 25,100 kcal/ha/day in 2000).

Moreover, yield disparities continued to grow: whereas the most productive region's yields were double those of the least productive regions in 1961, in 2000 they were 3.4 times greater.

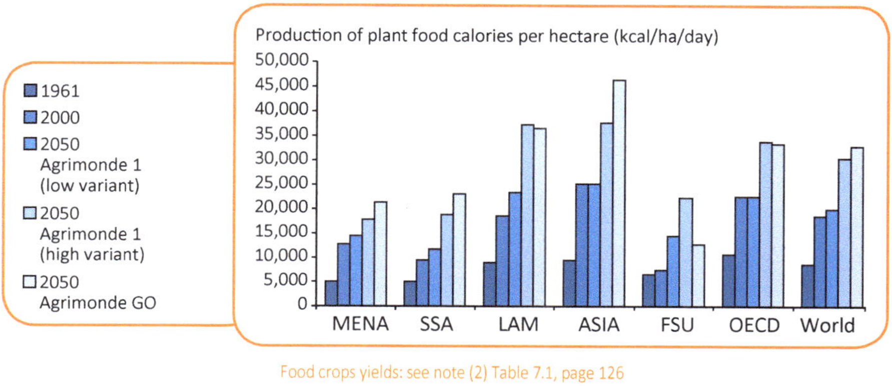

Fig. 7.7 Food crops yields in 1961 and 2000 and in 2050 in the Agrimonde scenarios. (Food crops yields: see note b Table 7.1, p. 126)

In Agrimonde GO, these two groups continue to exist until 2050. ASIA, whose food crop yields have grown by 85 % between 2000 and 2050, remains by far the most productive region in the world in 2050 with a yield level of around 46,000 plant kcal/ha/day in 2050. Even if OECD remains the third most productive region, its yield increases have been the lowest in the world (+48 %). In contrast, SSA is the region with the most spectacular yield gains (+144 %). In 2050, with MENA, it attains the yield levels of OECD in 2000. Finally, with a 71 % yield growth between 2000 and 2050, FSU remains the least productive region in the world. In this scenario, the gap between the least productive and the most productive regions widens slightly compared to 2000, ranging from 1 to 3.6 times greater in 2050 (Table 7.1 and Fig. 7.7).

In the Agrimonde 1 scenario, ASIA, OECD and MENA are the regions where few reserves for yield increases have been identified. Their food crop yields have therefore grown little over the period from 2000 to 2050 (0 to 13 % for the low variant of yields, and 40 to 50 % for the high variant). LAM and SSA have found it easier to improve their food crop yields. In this scenario they have followed the same progress (their yields have increased by respectively 30 % and 25 % in the low variant and by 100 % in the high variant). Finally, FSU has caught up spectacularly by doubling its food crop yield levels in the low variant, and tripling them in the high variant (Table 7.1, Figs. 7.5 and 7.7).

The two groups of regions differentiated by their yield levels in the past have changed little between 2000 and 2050. In both the low and high variant, in 2050 ASIA remains the most productive region in the world with a yield between 25,100 and 37,700 plant kcal/ha/day. Finally, the gap in yields between the most productive and the least productive region has shrunk. In 2050 the ratio is 1:2, i.e. its 1961 level (Table 7.2).

Table 7.2 Driving forces behind the evolution of food crop yields in Agrimonde 1

Region	Observed yields and driving forces of increasing yields	Forces limiting the increase of yields
MENA	Very rapid increase of yields over the period 1961–2003, without perceptible deceleration	Farming marginal lands
		Accentuation of water stress due to climate change
		Slowdown of yield gains in other foresight and forecasting exercises compared to past trends
SSA	Relatively low yield levels in 2000	Aridification due to climate change
	Continued increase of yields in other foresight or forecasting exercises	Deficit from the past in human capital and infrastructure providing access to markets
	Doubling of irrigated areas	
	Investments in research, training and agricultural development	
LAM	Continued increase of yields in other foresight and forecasting exercises	Fragility of the Cerrado soil and soil on the pioneer front in forests
	Heavy investments in the past in research, training and agricultural development	Aridification in Central America and of the Cerrado due to climate change
		Unequal access to factors of production
ASIA	Very fast increase of yields in the period 1961–1990	Farming of marginal lands
		Stagnation of yields per hectare since the early 1990s
		Slowdown of yield increases in other foresight and forecasting exercises
		Impacts of climate change: salinisation of irrigated rice farming deltas; increasingly frequent violent climatic events and accentuation of water stress in Northern China
FSU	Yield level relatively low in 2000	Slow increase of yields in other foresight and forecasting exercises
OECD	Very fast increase in yields in the period 1961–2000	Stagnation of yields per hectare since the early 2000s
		Slower yield increases in other foresight and forecasting exercises
		Agricultural sector objectives moving towards higher quality in food products

Chapter 8
Resource-Use Balances in the Agrimonde Scenarios

Tristan Le Cotty and Jean-Marc Chaumet

In Agrimonde 1, the quantitative assumptions on resource production in each region and the assumptions on resource consumption were established independently, in relation to past trends and to factors that the panel considered to be drivers for the future (Chaps. 5, 6 and 7). To obtain quantitative scenarios, it is therefore essential to check whether these assumptions are compatible with one another and under what conditions of trade between regions, and then, if necessary, to adjust the values of certain assumptions.

To perform this test of coherence, we review resource-use balances per region, according to the same principle used in the retrospective analysis (Chaps. 2 and 3). The specific nature of the 2050 resource-use balances is that they contain certain unknowns which must be deduced from the adjustment process (to reach an equilibrium), especially the quantities of plant calories used for animal production. The rules of priority between the various possible uses of biomass are then laid down, in accordance with the philosophy of the scenario under consideration. For example, when a region has a shortage of animal calories, will it import plant calories to feed its livestock, or rather directly import the animal calories that it lacks? From which regions? Such rules of balancing resources and uses of biomass are not trade simulations but rather a global test of coherence of the quantitative assumptions on the various scenario components.

This chapter begins with a description of the resource-use balance of the reference year, 2003. It then explains how a balance is achieved in a scenario, and finally it reviews the resource-use balance of both the Agrimonde 1 and Agrimonde GO scenarios.

T. L. Cotty (✉)
CIRAD—Umr CIRED, TA C56/15, 73 rue J.F. Breton,
34398, Montpellier Cedex 5, France
e-mail: lecotty@cirad.fr

J.-M. Chaumet
Institut de l'Elevage, 147 Rue de Bercy,
75595 Paris Cedex 12, France
e-mail: Jean-Marc.Chaumet@inst-elevage.asso.fr

S. Paillard et al. (eds.), *Agrimonde – Scenarios and Challenges for Feeding the World in 2050,* 123
DOI 10.1007/978-94-017-8745-1_8, © Éditions Quæ, 2014

Table 8.1 Food resource-use balance in 2003

Region		Use (Gkcal/day)	Resource (Gkcal/day)	Balance (Gkcal/day)
MENA	Plant	1,985	1,262	−724
	Animal	153	126	−28
SSA	Plant	2,191	1,938	−253
	Animal	108	93	−15
LAM	Plant	3,109	3,766	+657
	Animal	394	377	−18
ASIA	Plant	11,807	11,647	−160
	Animal	1,230	1,186	−44
FSU	Plant	1,580	1,619	+40
	Animal	220	197	−23
OECD	Plant	8,669	9,109	+440
	Animal	1,437	1,565	+127
World	Plant	29,341	29,341	0
	Animal	3,544	3,543	0

The Resource-Use Balance in 2003

The 2003 resource-use balance illustrates the global situation at the beginning of the twenty-first century. A summarised version of this balance is presented in Table 8.1, in which all the uses of biomass are grouped together in the "use" column.

Only calories from land sources are represented here, and are divided into two categories: plant calories and animal calories.

Uses include the calories available for human consumption[1], as well as calories used for animal feed, seed, non-food uses, and waste[2] (Fig. 8.1). Resources include regional production and the calories traded with other regions.

The balance for the year 2003 shown in Table 8.1 serves to measure the surplus or deficit of plant and animal calories for food at regional level. Three regions are thus shown to have a deficit of both plant and animal calories: North Africa—Middle East (MENA), where net imports cover 36 % of the use of plant calories and 18 % of the use of animal calories; sub-Saharan Africa (SSA), where they cover 12 % of the use of plant calories and 14 % of the use of animal calories; and ASIA, with 1 % and 4 %, respectively. Total net imports of these three regions account for 4 % of global plant production and 2 % of global animal production. Two regions have a surplus of plant calories and a deficit of animal calories: Latin America (LAM) (net exports of plant calories account for 17 % of the region's production and net imports of animal calories account for 5 % of consumption), and the former Soviet

[1] Apparent food availability is the calorie equivalent of the quantities of food available for the inhabitants of a region, calculated as follows: production + imports—exports +/-stock variations-animal feed-non-food uses-seed-waste after harvesting (Chap. 2). It includes waste after food reaches consumers (households or collective catering).

[2] Loss before food reaches consumers (Chaps. 2 and 5, Appendix 3, p. 261).

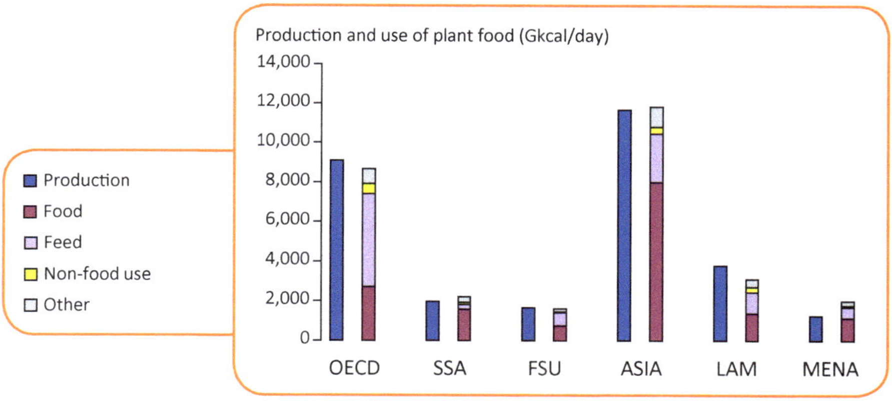

Fig. 8.1 Production and use of plant food in 2003 (Gkcal/day). (Source: Based on Agribiom, FAO, MA, and Agrimonde)

Union (FSU) (net exports of plant calories account for 2 % of the production and net imports of animal calories account for 10 % of consumption). Only the OECD-1990 (OECD) has a surplus of plant and animal calories (net exports account for 5 % of the production of plant calories and 8 % of the production of animal calories).

Achieving a Resource-Use Balance in the Scenarios

In many models, consumers' and producers' reactions to price variations, namely price elasticities, are assumed to be known, and the quantities consumed, produced and traded are derived from the economic equilibrium, as are the levels of economic welfare attained. In the Agribiom simulation tool, the levels of consumption, production and trade are deduced from a series of assumptions discussed and decided by the foresight expert panel, and from the rules enabling the panel to complete its assumptions, to test their coherence and, where relevant, to make adjustments.

Exogenous and Endogenous Variables

For food biomass resources, the exogenous variables (their value being settled by the panel) are firstly areas under food crops and their productivity in terms of food calories (Chaps. 6 and 7). These serve to calculate the regional production of plant calories for food. Assumptions are also made on the areas under pastures, taken into account in the calculation of food resources of animal origin. All the other variables concerning resources are deduced from the resource-use balance: animal calories produced, and trade balance in plant and animal calories.

The exogenous variables of food calorie uses are primarily the animal and plant calories assumed to be available for human consumption (in Gkcal/cap/day). They are obtained, at regional level, by multiplying the population by the mean availability of plant and animal calories (in kcal/cap/day). The other exogenous variables concern seed calories, calories used for non-food purposes (biofuels, etc.), animal calories used in animal production (milk powder or other), and calories lost before reaching consumers. These variables are calculated on the basis of a set percentage of total calories used in a given region. Finally, the quantity of plant calories used for animal feed is deduced from the functions of animal production during the adjustment process, and therefore depends on the rules chosen.

Rules for Obtaining Resource-Use Balances

There are several ways of obtaining a global balance, notably in relation to the quantity of plant calories used in each region for animal production. We therefore set rules determining priorities between the various competing uses of the available calories (human food, animal feed, biofuels, etc.), in order to describe only one balance for each variant of the scenario. In Agrimonde, these rules were essentially on the nature (animal or plant origin) and regional origin of the calories traded. These rules also constituted assumptions on the future, which should be borne in mind when interpreting the balances of the scenarios.

To illustrate the necessity to set such adjustment rules, we can imagine a region in which plant production does not suffice to cover both human consumption of plant products and the use of plants to produce animal products consumed in the same region. In such cases of deficit, it is possible to balance the region's resources and uses by favouring either plant calorie imports for the production of the required animal products in the region, or else direct animal calorie imports. The transformation of plant calories into animal calories is not the same in these two cases, because each region has specific breeding systems and therefore different functions of animal production. The global balance obtained is not the same when the regions trade plant calories and when they trade animal calories. In fact, both types of product are actually traded, but simulating these two extreme cases provides an indication of the diversity of the possible adjustment conditions.

In Agrimonde, the two variants of adjustment rules were used: the first was based on trade in plant calories and the second on trade in animal calories. The two adjustment variants are therefore two modes of calculation corresponding to two possible conditions of trade in calories. They do not enable us to simulate trade in 2050 as if there were equilibrium prices; they simply illustrate how the nature and origin of the calories traded can influence the balance. The purpose of calculating balances is therefore to verify whether the consumption assumptions are compatible with the production assumptions under these two variants, with each variant constituting a test of coherence of the scenarios.

In Variant 1, priority is given to trade in plant calories. Each region produces animal calories to meet its exact needs. To do so, it has specific regional technology relating to animal production, pasture areas, and a certain quantity of plant calories (grains, etc.) for animal feed, which are produced regionally and, if necessary, imported. The functions of animal production make it possible to calculate the quantity of plant calories needed in each region for animal feed, in order to attain the right level of animal production to meet the region's needs (Appendix 5, p. 270). All the region's uses of plant calories therefore include those required to cover its needs in animal calories. The region's trade balance, expressed in plant calories, is obtained by calculating the difference between its plant production and all its uses. On a global scale, if the sum of these regional balances is positive, then the global resources exceed the resources needed to satisfy the consumption assumptions; if it is negative, then they are insufficient, which makes it necessary to adjust the assumptions on uses and/or on resources.

In Variant 2, priority is given to trade in animal products. Each region uses its plant production primarily for human food, non-food uses and seed, and experiences waste after harvesting. If it has plant calories left over, it uses them for animal feed. The quantity of animal calories produced is given by the functions of animal production. The regional trade balance is therefore the difference between the production of animal calories, calculated on the basis of the functions of animal production, and the total regional uses of animal calories. A region can find itself in a situation of having to import not only animal calories but also plant calories when its plant production does not cover its regional needs in plants for human food. In that case, the scenario must also include rules specifying which regions export plants and in what proportions (Appendix 5, p. 270).

Variant 1 therefore yields regional balances in a single unit (plant calories) which take into account both human food and animal feed in each region. Variant 2 presents the trade balances in two units: plant calories and animal calories, as both types of calorie are intended for human consumption.

Global Resource-Use Balance in the Agrimonde GO Scenario

Recap of the Agrimonde GO Assumptions

The Agrimonde GO scenario is inspired by the Global Orchestration scenario of the Millennium Ecosystem Assessment (MA) (MA 2005b). It applies most of the MA's quantitative assumptions for the Global Orchestration scenario, although the adjustment process for reaching a resource-use balance in Agrimonde GO is that designed for Agribiom. This is to some extent a trend-based scenario of each region's resources and uses, showing sustained growth of yields and of consumption in most regions of the world up to 2050. The main assumptions of this scenario are shown in Table 8.2.

The population assumption is not the same as that of the MA; it corresponds to the United Nations' median population assumption for 2050, which was also applied in the Agrimonde 1 scenario (Chap. 5). Total food availability per region,

Table 8.2 Uses and resources in 2003 and main assumptions of the Agrimonde GO scenario in 2050

Region	Population (M inhabitants)		Total supply (kcal/cap/day)		Cropland[a] (Mha)		Pasture[b] (Mha)		Yield (Gkcal/ha/day)	
	2003	2050	2003	2050	2003	2050	2003	2050	2003	2050
MENA	372	632	3,340	3,457	84	93	328	320	15,010	21,362
SSA	706	1,662	2,353	2,972	202	303	784	1,161	9,582	23,133
LAM	538	774	3,125	3,698	164	266	553	548	22,979	36,493
ASIA	3,322	4,427	2,762	3,702	461	504	565	735	25,251	46,416
FSU	279	239	3,250	3,457	202	223	360	212	8,026	12,825
OECD	987	1,066	3,908	4,099	416	467	736	608	21,904	33,507

[a] The quantitative assumptions and results are all presented in Appendix 5, p. 265, as are the values of the corresponding variables recorded in 1961 and 2003

[b] The surface areas in Agrimonde GO are those of the Global Orchestration scenario of the MA, adjusted to take into account the differences in the definition of land use according to the FAO and the MA (Chap. 6, Appendix 4, p. 263)

which reflects the region's food consumption, demonstrates a continuation of the growth recorded from 1961 to 2003. It is adapted from the Global Orchestration scenario of the MA, whose assumptions for food consumption concerned only cereals and meat (from grazing and non-grazing animals). The assumptions on the availability of human food in Agrimonde GO have thus been calculated by applying the growth rate projected in Global Orchestration for cereal consumption to all human consumption of plant calories. Likewise, by applying the growth rate of meat consumption in Global Orchestration to total animal calories, assumptions were obtained in Agrimonde GO on animal calorie availability (meat from non-grazing and grazing animals, as well as eggs and dairy products) (Appendix 3, p. 261).

Assumptions on surface areas (croplands and pastures) are based on moderate growth which differs little from the 1961–2003 trend. They are based on the assumptions of Global Orchestration, with slight adjustments to correct the discrepancy in the total surface area between MA data and FAO data, and to take into account the countries excluded from Agribiom (Appendix 4, p. 263). The yield assumptions slightly resemble a continuation of 1961–2003 growth, resulting from rapid technological progress. In the MA scenarios the assumptions on yields concern only cereals. The Agrimonde GO assumptions on plant calorie yields were thus obtained by applying the growth rate of cereal yields in the Global Orchestration scenario to all plant production. Finally, seed, non-food use, and loss between production and availability to consumers are assumed to be identical in 2003 and 2050, in terms of the proportion of total use.

The Resource-Use Balance in Variant 1 of Agrimonde GO

The balance in the Agrimonde GO scenario, summarised in Table 8.3 and Fig. 8.2, gives an overview of the regional assumptions for the different variables. The first conclusion is that biomass resources cover uses at global level, which tends to confirm the coherence in terms of meeting food needs with resources in the Global Orchestration scenario of the MA.

Table 8.3 Food resource-use balance in 2050, Agrimonde GO scenario, Variant 1

Region		Use (Gkcal/day)	Production (Gkcal/day)	Net trade (Gkcal/day)
MENA	Plant	4,176	1,985	−2,190
	Animal	335	335	0
SSA	Plant	7,378	6,084	−1,294
	Animal	528	528	0
LAM	Plant	5,930	7,992	+2,062
	Animal	835	834	0
ASIA	Plant	23,009	22,094	−915
	Animal	4,188	4,189	0
FSU	Plant	2,118	2,398	+280
	Animal	363	363	0
OECD	Plant	10,939	13,436	+2,497
	Animal	2,159	2,158	0
World	Plant	53,551	53,990	+440
	Animal	8,408	8,407	0

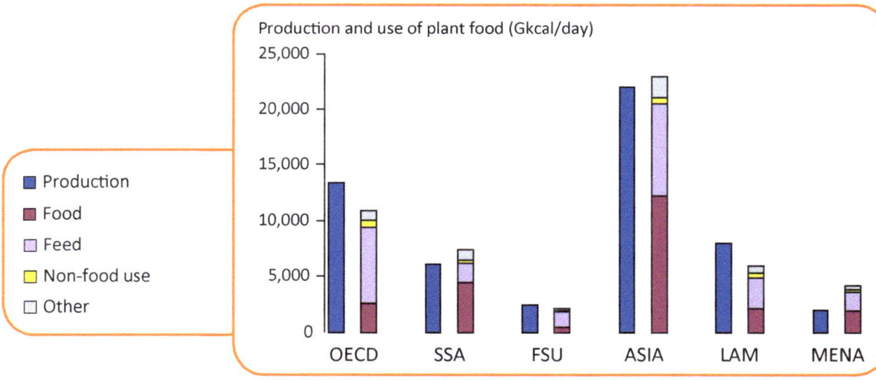

Fig. 8.2 Production and use of plant food in 2050, Agrimonde GO scenario, Variant 1 (Gkcal/day). (Source: Based on Agribiom, FAO, MA, and Agrimonde)

In the first simulation variant, by construction, the animal calorie resource-use balances are nil in all the regions, since each region produces the animal biomass required to meet its needs, either by using its plant production, or by importing plant calories for animal feed. The balances thus apply only to plant calories and include the need to import to meet all needs, both human and animal.

The regional assumptions and adjustment process selected tend to accentuate net regional deficits or surpluses compared to 2003. Thus, the three regions that already had deficits in 2003 have even greater deficits in 2050: by decreasing order of magnitude of the deficit, MENA, SSA and, lastly, ASIA. In MENA, crop production is not enough to meet even human food needs (this does not appear in Table 8.3 which clusters uses). Three regions have a surplus: by ascending order of magnitude FSU, LAM, OECD.

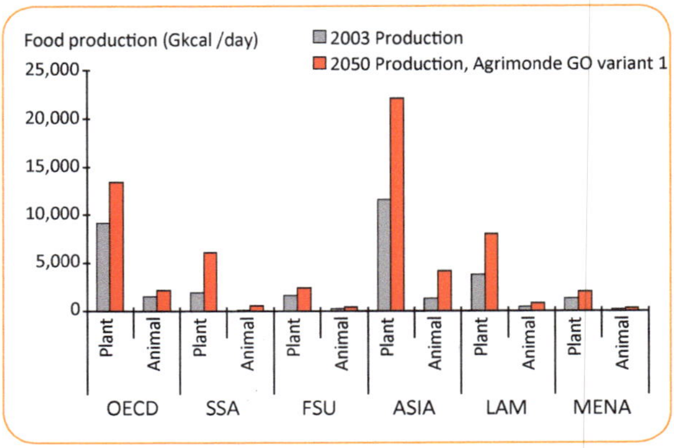

Fig. 8.3 Food production in 2003 and 2050, Agrimonde GO scenario, Variant 1. (Source: Based on Agribiom, FAO, MA, and Agrimonde)

In this scenario, between 2003 and 2050, plant production increases by 84% and animal production by 137%. This growth affects all the regions and products (Fig. 8.3) and all uses (Fig. 8.4), and the regional disparities in the per capita uses of calories are maintained (Fig. 8.5). Finally, Fig. 8.6 shows the continuous growth of per capita mean global consumption of plant and animal calories in the Agrimonde GO scenario.

The Resource-Use Balance in Variant 2 of Agrimonde GO

In Variant 2, the net trade balance includes the plant and animal calories required to meet human needs in both types of calorie. The global balance remains close to an equilibrium, and the same three regions show deficits (Table 8.4). On the other

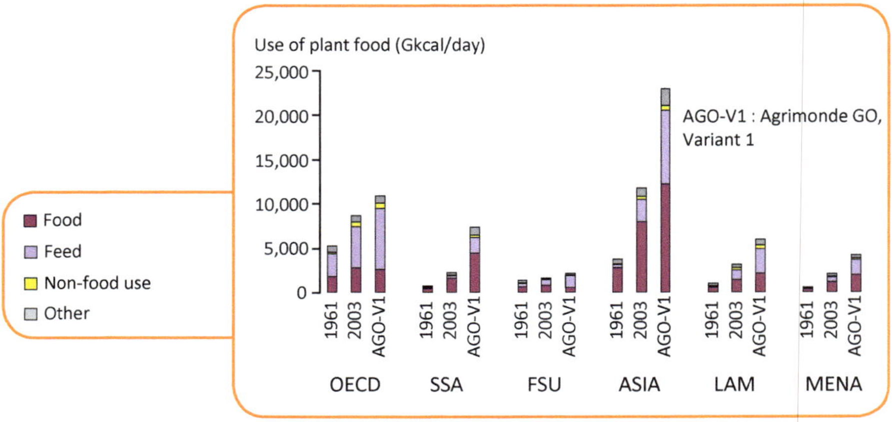

Fig. 8.4 Use of plant food in 1961, 2003 and 2050, Agrimonde GO scenario, Variant 1. (Source: Based on Agribiom, FAO, MA, and Agrimonde)

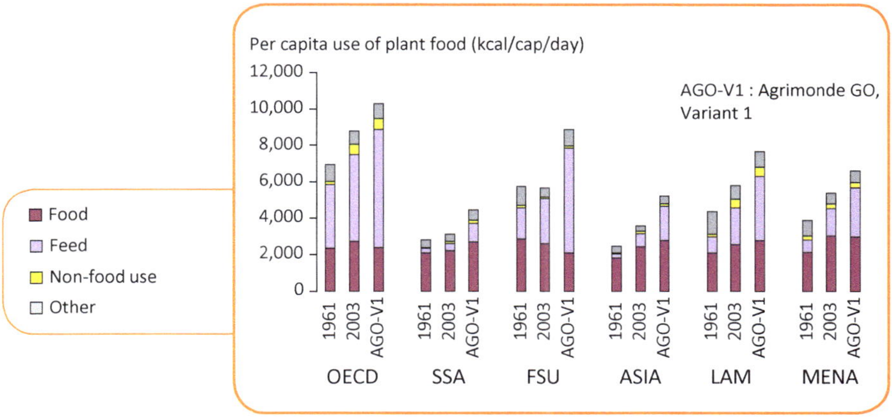

Fig. 8.5 Per capita use of plant food in 1961, 2003 and 2050, regional averages, Agrimonde GO scenario, Variant 1. (Source: Based on Agribiom, FAO, MA, and Agrimonde)

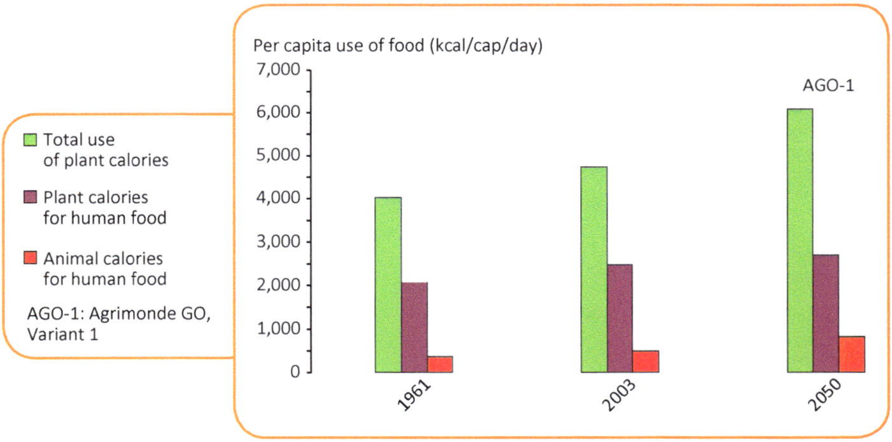

Fig. 8.6 Per capita use of food in 1961, 2003 and 2050, world average, Agrimonde GO scenario, Variant 1. (Source: Based on Agribiom, FAO, MA, and Agrimonde)

hand, uses of plant calories are very different, since the regions with deficits use (and import) fewer plant calories, and produce fewer animal calories, which they import in greater quantities. Since the transformation of plant calories into animal calories does not take place in the same regions as in Variant 1, and since the modes of animal production differ from one region to the next, the global balance is different (Fig. 8.7).

Here, the global surplus is expressed in animal calories, but it can be compared to the surplus obtained in Variant 1 by calculating the plant calories theoretically needed to obtain this surplus of animal calories. By using, for example, the production function of OECD, the global balance in animal calories of 60 Gkcal/cap/day

Table 8.4 Food resource-use balance in 2050, Agrimonde GO scenario, Variant 2

Region		Use (Gkcal/day)	Production (Gkcal/day)	Net trade (Gkcal/day)
MENA	Plant	2,223	1,985	−238
	Animal	335	25	−309
SSA	Plant	6,084	6,084	0
	Animal	528	219	−309
LAM	Plant	7,913	7,992	+79
	Animal	836	1,301	466
ASIA	Plant	22,094	22,094	0
	Animal	4,192	3,780	−412
FSU	Plant	2,319	2,398	+79
	Animal	363	403	+40
OECD	Plant x	13,357	13,436	+79
	Animal	2,154	2,738	+584
World	Plant	53,990	53,990	0
	Animal	8,407	8,466	+60

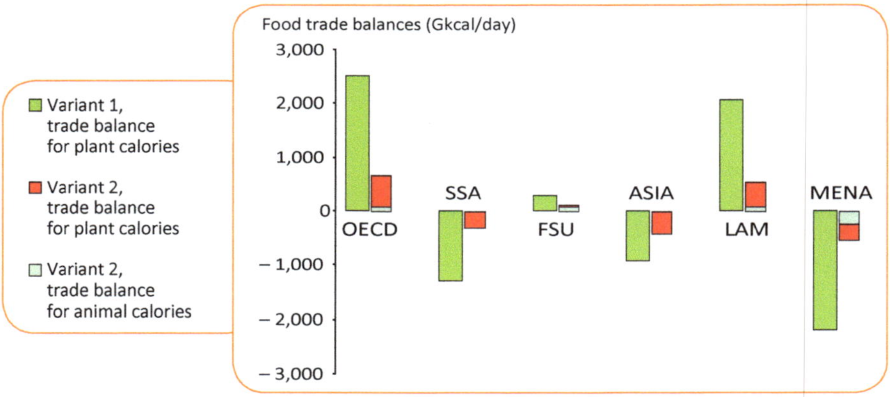

Fig. 8.7 Regional food trade balances, Agrimonde GO scenario. (Source: Based on Agribiom, FAO, MA, and Agrimonde)

represents 230 Gkcal/day[3] in plant calories, which is considerably lower than the 440 Gkcal/day of the global balance in Variant 1.

Variant 1 thus appears to have a greater surplus than Variant 2. This means that the same level of consumption can be attained from a lower level of plant production, depending on the place in which plant calories are transformed into animal calories. In the Agrimonde GO scenario—which cannot be generalised—the global balance has a larger surplus when the countries with deficits import plant calories to feed their livestock, than when they directly import animal calories. This result applies only to the quantitative assumptions chosen and the production functions used.

[3] This figure denotes the animal feed needed in the OECD region to produce 60 Gkcal/day of animal products.

Table 8.5 Uses and resources in 2003 and main assumptions of the Agrimonde 1 scenario in 2050

Region	Population (Million inhabitants)		Total supply (kcal/cap/day)		Cropland (Mha)		Pasture(Mha)		Yield (Gkcal/ha/day)	
	2003	2050	2003	2050	2003	2050	2003	2050	2003	2050
MENA	372	632	3,340	3,000	84	90	328	321	15,010	14,500
SSA	706	1,662	2,353	3,000	202	339	784	691	9,582	11,750
LAM	538	774	3,125	3,000	164	310	553	445	22,979	23,500
ASIA	3,322	4,427	2,762	3,000	461	560	565	512	25,251	25,100
FSU	279	239	3,250	3,000	202	310	360	300	8,026	14,500
OECD	987	1,066	3,908	3,000	416	495	736	576	21,904	22,600

The Resource-Use Balance in Agrimonde 1

Recap of the Assumptions of the Agrimonde 1 Scenario

The Agrimonde 1 scenario is based on assumptions that differ from past trends, recapitulated in Table 8.5. From the point of view of consumption, the originality of the Agrimonde 1 scenario is that it foresees a mean availability of food, in terms of calories, that is identical in all the regions. This implies substantial increases in certain regions and equally substantial decreases in others. The assumption of global consumption is on the whole far lower than in the Agrimonde GO scenario, and slightly higher than in 2003, due simply to the population increase (as Agrimonde 1 per capita global consumption is more or less the same as in 2003).

From the point of view of production, the scenario is based on assumptions of more extensive growth of cultivated areas than in Agrimonde GO. The expert panel defined a range of yields for each region. As the assumptions corresponding to the bottom end of the range enabled us to achieve a global balance for Variant 1, we selected these assumptions for Agrimonde 1[4]. The yield assumptions of this scenario are far lower than those in Agrimonde GO.

The Resource-Use Balance in Variant 1 of the Agrimonde 1 Scenario

The nil global balance (Table 8.6) shows that the consumption assumptions in the Agrimonde 1 scenario can be satisfied with the production assumptions, and with the adjustment rules retained in Variant 1.

In 2050, global production of plant calories in Agrimonde 1 has increased by close to 30 % since 2003, and the world produces 20 % more animal calories (Fig. 8.9). The regions with shortages are, in descending order: MENA, ASIA, and SSA. In SSA and

[4] In the calculations of resource-use balances, the low variants proposed by the panel produced a near balance in Variant 1. They were therefore marginally adjusted to obtain a precise global resource-use balance in this variant. Thus, the low variants presented in Chap. 7 and in the present chapter are not exactly the same as those initially formulated by the panel, but they are very similar.

Table 8.6 Food resource-use balance in 2050, Agrimonde 1 scenario, Variant 1

Region		Use (Gkcal/day)	Production (Gkcal/day)	Net trade (Gkcal/day)
MENA	Plant	3,549	1,302	−2,247
	Animal	335	335	0
SSA	Plant	7,515	3,525	−3,990
	Animal	852	852	0
LAM	Plant	3,977	5,875	+1,898
	Animal	431	431	0
ASIA	Plant	16,732	13,554	−3,178
	Animal	1,918	1 918	0
FSU	Plant	1,017	4,350	+3,333
	Animal	106	106	0
OECD	Plant	4,856	9,040	+4,184
	Animal	632	632	0
World	Plant	37,646	37,646	0
	Animal	4,274	4,274	0

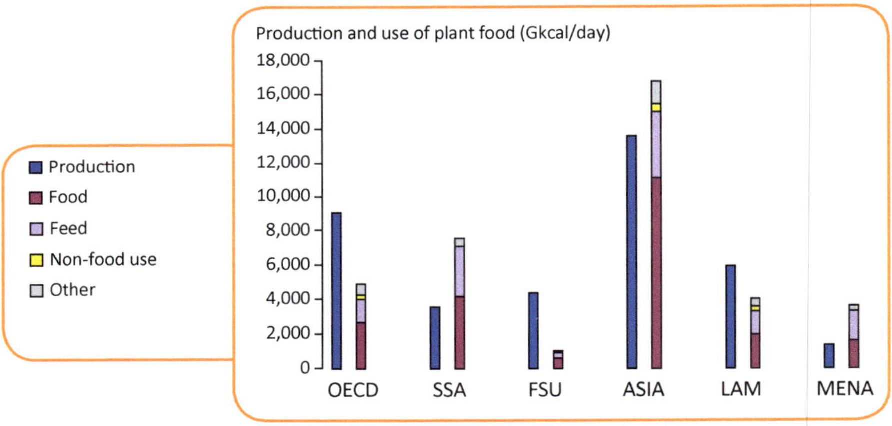

Fig. 8.8 Production and use of plant food in 2050, Agrimonde 1 scenario, Variant 1. (Source: Based on Agribiom, FAO, MA, and Agrimonde)

MENA, regional production of plant calories is not enough to satisfy human food needs. Some of these regions' imports (16 and 12%, respectively) are intended directly for human food, while the rest is used for animal feed. Plant product resources in ASIA are sufficient to meet direct human food needs, and some animal feed needs.

ASIA therefore has to import in order to offset the shortfall of calories for animal feed and to meet its animal calorie consumption needs. Regions with surpluses are, in descending order: OECD, FSU, and LAM.

As shown in Fig. 8.8, in the Agrimonde 1 scenario, the regional gaps between production and consumption are relatively large (larger than in Agrimonde GO). OECD and especially FSU have increasing surpluses due to the considerable

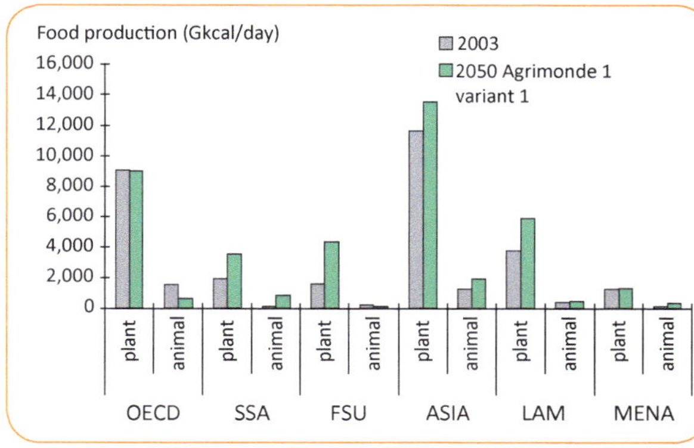

Fig. 8.9 Food production in 2003 and 2050, Agrimonde 1 scenario, Variant 1. (Source: Based on Agribiom, FAO, MA, and Agrimonde)

decline in consumption (resulting from the lower quantities of calories ingested and/or wasted after food reaches consumers). In contrast, SSA has a growing deficit due to a steep rise in consumption that local production cannot satisfy (assumption of yield increases far lower than those in Agrimonde GO). Finally, ASIA's deficit increase, notably due to saturation of its production capacities, is foreseen in Agrimonde 1, with regard to both surface areas and yields.

Compared to 2003, Figs. 8.10 and 8.11 also show that essential change lies in the reduced use of plant calories for animal feed. Thus, OECD shifts from a plant calorie total use of close to 9,000 kcal/cap/day in 2003 to about 4,500 in 2050, without reducing its food consumption in plant calories but by reducing its consumption of animal calories by 700 kcal/cap/day.

Variant 1 of the Agrimonde 1 scenario shows that it is possible for the world to maintain its current mean per capita consumption by using fewer plant calories (Fig. 8.12). Two conditions suffice: reducing consumer waste, and partially relocating animal production to areas where the average rate of transformation is higher.

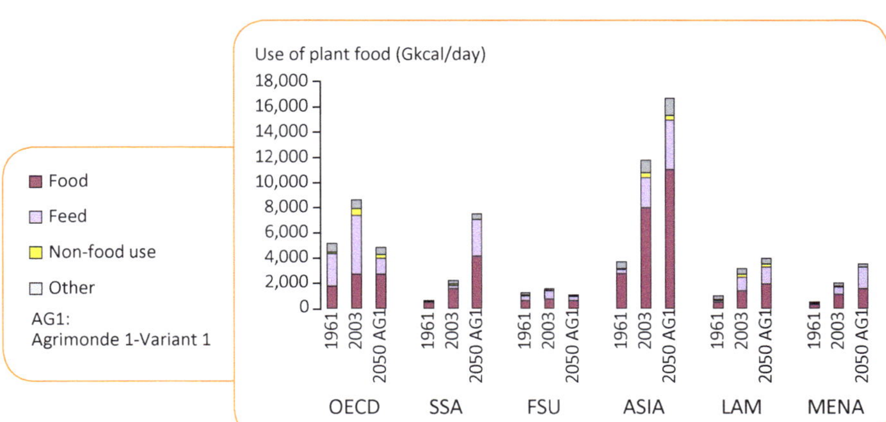

Fig. 8.10 Use of plant food in 1961, 2003 and 2050, Agrimonde 1 scenario, Variant 1. (Source: Based on Agribiom, FAO, MA, and Agrimonde)

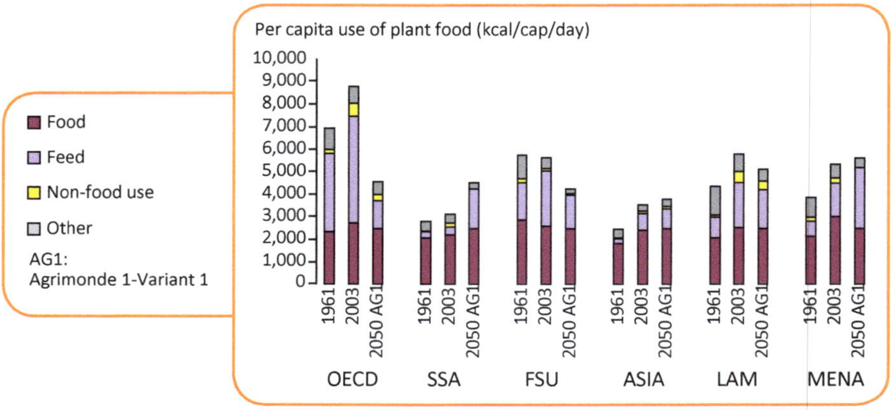

Fig. 8.11 Per capita use of plant food in 1961, 2003 and 2050, regional averages, Agrimonde 1 scenario, Variant 1. (Source: Based on Agribiom, FAO, MA, and Agrimonde)

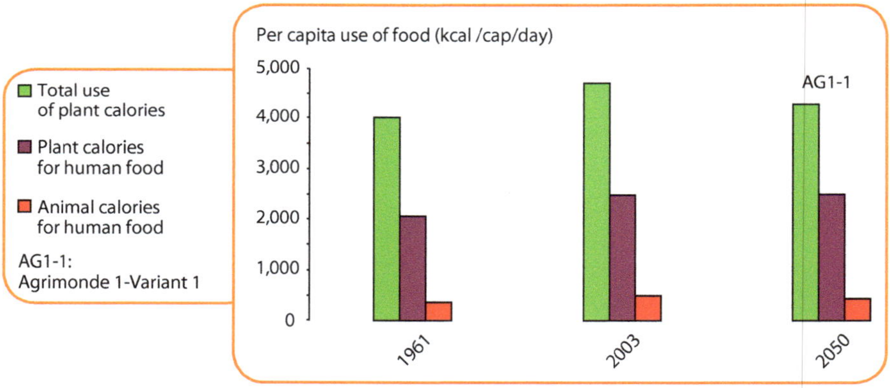

Fig. 8.12 Per capita use of food in 1961, 2003 and 2050, world average, Agrimonde 1 scenario, Variant 1. (Source: Based on Agribiom, FAO, MA, and Agrimonde)

The Resource-Use Balance in Variant 2 of the Agrimonde 1 Scenario

In Variant 2, with the same quantitative assumptions, the global resource-use balance is no longer nil but negative[5] (Table 8.7). In this variant, in which the regions directly trade animal calories rather than plant calories for animal feed, we can conclude that the Agrimonde 1 assumptions on resources are insufficiently high to cover the uses foreseen in the same scenario.

[5] It is possible to achieve a global balance by altering the quantitative assumptions (e.g. yields), but we chose to work with unchanged quantitative assumptions and to note (and interpret) the effect of the adjustment process on the global and regional balances.

Table 8.7 Food resource-use balance in 2050, Agrimonde 1 scenario, Variant 2

Region		Use (Gkcal/day)	Production (Gkcal/day)	Net trade (Gkcal/day)
MENA	Plant	1,752	1,302	−450
	Animal	336	24	−311
SSA	Plant	4,426	3,525	−901
	Animal	851	22	−829
LAM	Plant	5,425	5,875	+450
	Animal	431	777	+346
ASIA	Plant	13,554	13,554	0
	Animal	1,918	481	−1,437
FSU	Plant	3,900	4,350	+450
	Animal	106	689	+584
OECD	Plant	8,590	9,040	+450
	Animal	631	1,525	+893
World	Plant	37,646	37,646	0
	Animal	4,273	3,519	−754

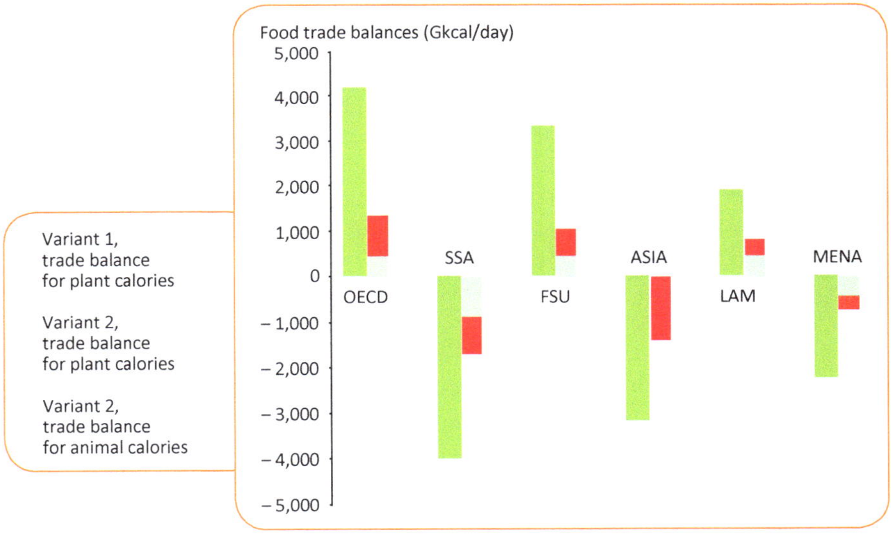

Fig. 8.13 Regional food trade balances, Agrimonde 1 scenario (Gkcal/day). (Source: Based on Agribiom, FAO, MA, and Agrimonde)

As in the Agrimonde GO scenario, Variant 2 seems to require more calories than Variant 1. The same three regions have deficits (Fig. 8.13): SSA, ASIA, and MENA. For the deficits of these regions to be offset by the surpluses of other regions (LAM, FSU and OECD), global production of animal calories would have to increase by 20 % (754 Gkcal/day).

Table 8.8 Food use in 2003 and 2050, Agrimonde GO and Agrimonde 1 scenarios (Gkcal/day)

Region		Use 2003 (Gkcal/day)	Use 2050 Agrimonde 1 Variant 1 (Gkcal/day)	Use 2050 Agrimonde G0 Variant 1 (Gkcal/day)
MENA	Plant	1,985	3,549	4,176
	Animal	153	335	335
SSA	Plant	2,191	7,515	7,378
	Animal	108	852	528
LAM	Plant	3,109	3,977	5,930
	Animal	394	431	835
ASIA	Plant	11,807	16,732	23,009
	Animal	1,230	1,918	4,188
FSU	Plant	1,580	1,017	2,118
	Animal	220	106	363
OECD	Plant	8,669	4,856	10,939
	Animal	1,437	632	2,159
World	Plant	29,341	37,646	53,551
	Animal	3,543	4,274	8,408

Conclusion

The comparison of a trend-based scenario like Agrimonde GO with the rupture scenario Agrimonde 1 (e.g. in Variant 1 on attaining a balance), illustrates two developments that are coherent in terms of covering uses with resources, and strongly contrasted in terms of agriculture and food. This coherence is established here owing to the resource-use balances in terms of food calories. Achieving a balance in these two variants provides a first test of the scenarios' sensitivity to preferences for trade in animal calories or plant calories.

The Agrimonde GO scenario shows steep growth of production compared to 2003, attaining 80 % for plant production and 140 % for animal production, with a human population growth of 42 %. Based on considerable yield increases, including in regions that already have intensive production, it provides all the regions of the world with an average level of availability greater than today's, and maintains major gaps in consumption between the regions (Table 8.8).

From the point of view of a global balance, Agrimonde 1 is a scenario with moderate production growth: approximately 30 % plant production growth and 20 % animal production growth in 2050 compared to 2003. Global population increases are equal to those in Agrimonde GO: 42 % (Figs. 8.14 and 8.15). Yet mean food availability increases in the regions where it is currently insufficient (i.e. less than the percapita availability targeted in this scenario): ASIA and SSA. The changes proposed in this scenario are therefore based not only on changes in production but also on a geographical redistribution of consumption.

Figure 8.16 illustrates the effect of the application of an average universal food ration on the uses of plant resources (3,000 kcal/cap/day in Agrimonde 1). It shows, in particular, the steep reduction in calories used for animal feed in OECD.

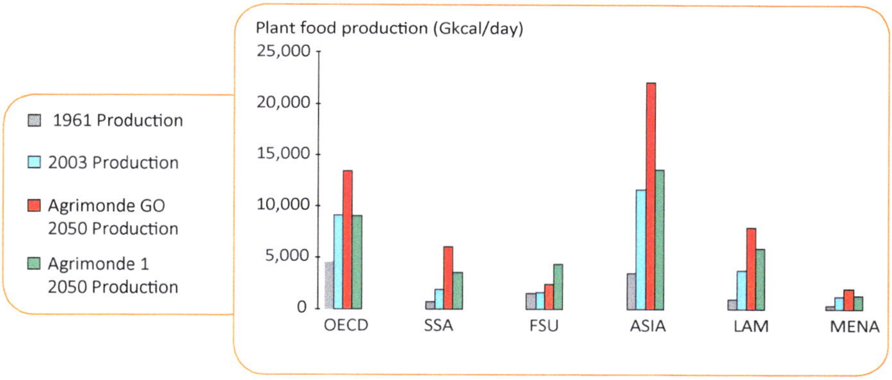

Fig. 8.14 Plant food production in 1961, 2003, 2050, Agrimonde GO and Agrimonde 1 scenarios (Gkcal/day). (Source: Based on Agribiom, FAO, MA, and Agrimonde)

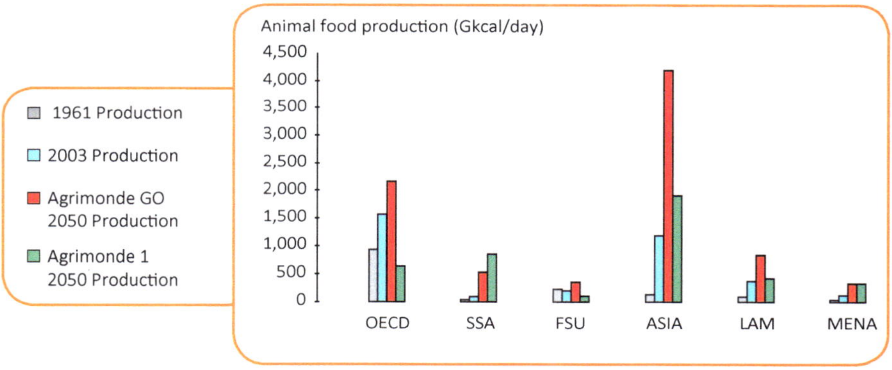

Fig. 8.15 Animal food production in 1961, 2003, 2050, Agrimonde GO and Agrimonde 1 scenarios. (Source: Based on Agribiom, FAO, MA, and Agrimonde)

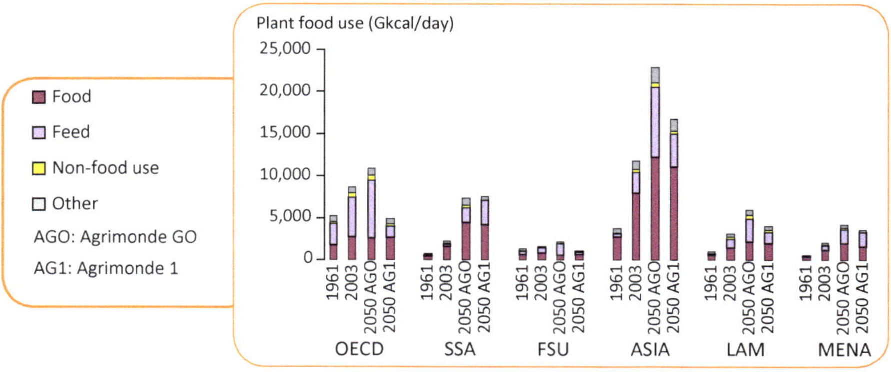

Fig. 8.16 Plant food use in 1961, 2003 and 2050, Agrimonde GO and Agrimonde 1 scenarios. (Source: Based on Agribiom, FAO, MA, and Agrimonde)

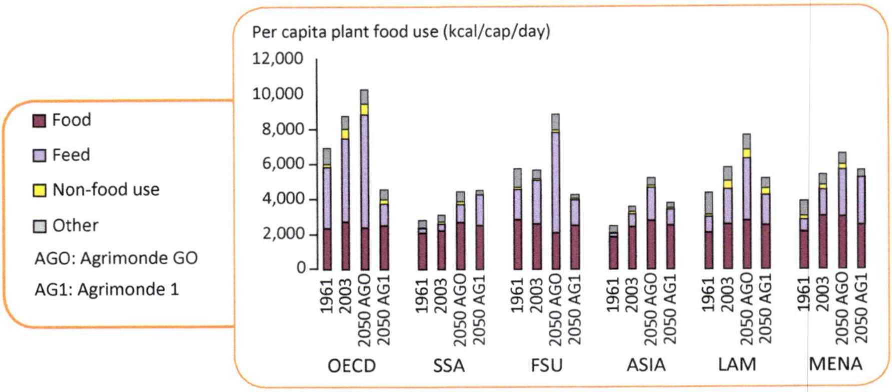

Fig. 8.17 Per capita plant food use in 1961, 2003 and 2050, regional averages, Agrimonde GO and Agrimonde 1 scenarios. (Source: Based on Agribiom, FAO, MA, and Agrimonde)

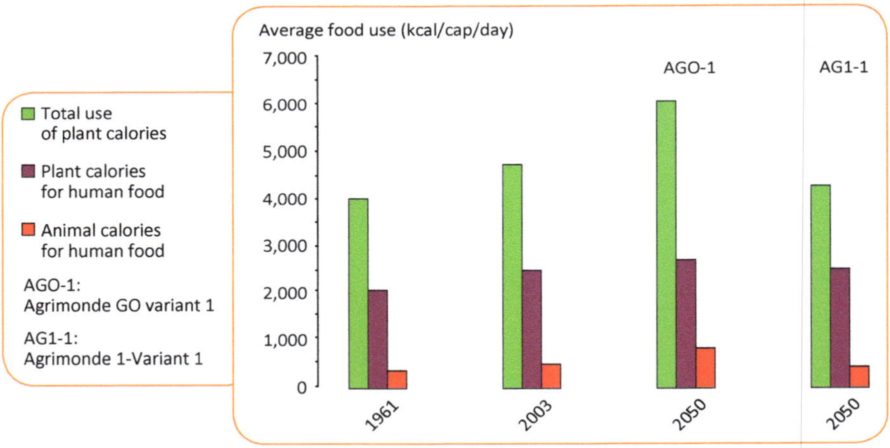

Fig. 8.18 Average food use in 1961, 2003 and 2050, world average, Agrimonde GO and Agrimonde 1 scenarios. (Source: Based on Agribiom, FAO, MA, and Agrimonde)

Figure 8.17 shows the same trend per capita. It illustrates the redistribution of calories in Agrimonde 1, compared to 2003 and to Agrimonde GO. Figure 8.18 presents an aggregated comparison of the two scenarios on a global scale, which enables us to distinguish total consumption of plant calories, consumption of plant calories available for human food, and consumption of animal calories for food.

Chapter 9
Agrimonde 1 and Agrimonde GO: Comparison, Coherence, Drivers of Change

Tévécia Ronzon, Jean-Marc Chaumet and Sandrine Paillard

This chapter examines the quantitative scenarios built, so as to define their qualitative dimensions, left undetermined by the quantitative analysis. Initially, for each region and globally, the scenarios were tested for their internal coherence and conclusions were drawn from their comparison. The aim was to establish whether, on both a regional and global scale, the quantitative assumptions for the different variables make it possible to put together truly coherent scenarios, that is, compatible with the principles chosen initially, particularly the principle of sustainability in the case of the Agrimonde 1 scenario. Comparison of the two scenarios (Agrimonde 1 and Agrimonde GO) enables us to oppose two very different strategies, both of which (theoretically) allow for a food resource-use balance on a global scale, as shown in Chap. 8.

In this respect, in this chapter and the following we refer to the resource-use balances calculated on the basis of Variant 1, i.e. the variant in which inter-regional trade is in the form of plant calories only (countries with a shortage of animal calories import plant calories to feed their livestock). The analysis then focuses on the conditions and drivers of change leading to the world of 2050 described in Agrimonde 1 in each region and globally (the conditions of Agrimonde GO are assumed to be those of the Global Orchestration scenario of the Millennium Ecosystem Assessment (MA) (MA 2005b)).

T. Ronzon (✉) · S. Paillard
IPTS, C/Inca Garcilaso, s/n 41092, Seville, Spain
e-mail: tevecia.ronzon@ec.europa.eu

S. Paillard
ANR, 212 rue de Bercy, 75012 Paris, France
e-mail: sandrine.paillard@agencerecherche.fr

J.-M. Chaumet
Institut de l' Elevage, 147 Rue de Bercy, 75595 Paris Cedex 12, France
e-mail: Jean-Marc.Chaumet@inst-elevage.asso.fr

S. Paillard et al. (eds.), *Agrimonde – Scenarios and Challenges for Feeding the World in 2050,*
DOI 10.1007/978-94-017-8745-1_9,

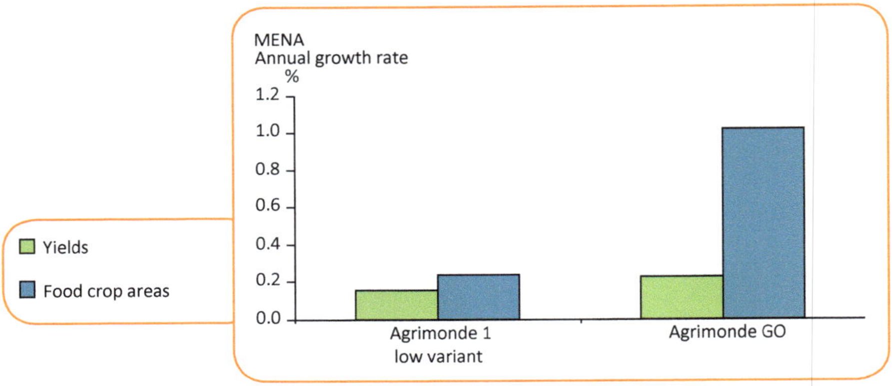

Fig. 9.1 MENA: Annual growth rate of food crop areas and yields in the Agrimonde scenarios between 2000 and 2050

Middle East—North Africa: The Challenge of Development Based on Increasingly Rare Natural Resources

In Middle East—North Africa (MENA) the population has multiplied by 1.8 between 2000 and 2050 in both scenarios. The per capita demand for calories has declined by 10 % in Agrimonde 1 and has remained more or less stable in Agrimonde GO (+3 %). The total food calorie demand (the daily per capita availability of calories of plant, animal and aquatic origin for human consumption, multiplied by the population) has dropped to 1,900 Gkcal/day in Agrimonde 1 and 2,200 Gkcal/day in Agrimonde GO. The composition of diet in terms of calorie origin is much the same in the two scenarios.

Between 2000 and 2050, the scarcity of water resources in this region has worsened due to climate change. This has limited the possibilities of increasing production, while the calorie demand has increased. Cultivated areas (food crop mainly) have changed little (+7 million ha: +8 %) and in 2050 they occupy the region's full cultivation potential. Pastures have decreased very little (−6 million ha) and food crop yields have increased by only 13 % in 50 years in the low variant (+0.24 % per annum). In Agrimonde GO, cultivated surfaces have increased by 10 million ha and pastures have lost 7 million ha; food crop yields have increased by 66 % (i.e. +1.05 % per annum, Fig. 9.1).

Total use of plant calories (the sum of use for human food, animal feed, seed, loss, and non-food uses) in MENA has risen from 1,900 Gkcal/day in 2000 to 3,500 in 2050 in the Agrimonde 1 scenario (+90 %) and to 4,200 Gkcal/day in Agrimonde GO (+125 %) (in Variant 1 of the resource-use balances of the scenarios, Chap. 8). In terms of resource-use balance, the strategies adopted in Agrimonde 1 and in Agrimonde GO do not enable MENA to meet its calorie demand in 2050. There is a shortage of about 2,200 Gkcal/day in both scenarios. Thus, the demand for plant calories for human consumption can be met but regional resources are insufficient to feed all the livestock.

Are the Agrimonde Scenarios Coherent in Middle East—North Africa?

The resource-use balances suggest the need to examine the coherence of the Agrimonde 1 scenario. In particular, as this region will probably not be able to rely on income from fossil fuels in 2050, the question arises of which economic sectors will be able to ensure that the food demand considered in the Agrimonde 1 scenario is met, assuming that the development of the agricultural sector is strongly limited. Moreover, is it really possible to maintain agricultural surfaces in a context of saturated cultivation potential combined with growing urbanisation[1]?

In the Agrimonde 1 scenario, the development of the agricultural sector is limited due to a lack of potential croplands and to environmental constraints, especially water, which have severely hindered the possibilities of increasing yields. Compared to the Agrimonde GO scenario, which assumes a considerable increase in yields that have allowed for gains in agricultural income, in Agrimonde 1 the agricultural sector may not be able to employ a fast-growing rural population. Given the extent of projected population growth in MENA, a large proportion of the rural population may migrate to urban areas. Our scenario is then faced with an essential problem of coherence since the limitation of cultivable land could result in an acceleration of the artificialisation of agricultural land[2], leading to a vicious circle that does not make it possible to maintain agricultural areas[3]. Moreover, the inequalities of living conditions between urban and agricultural populations could increase dangerously and constitute another problem of coherence for this scenario. The sustainability of the Agrimonde 1 scenario therefore relies heavily on opportunities for wealth creation in rural areas, through agricultural production with high added value, and in product processing, tourism, etc.

The problems and questions raised by the Agrimonde GO scenario seem to be much the same. Moreover, in the latter scenario the assumption of a sharp rise in yields is relatively strong since it implies that the impacts of climate change (e.g. greater variability of precipitations) have been overcome in this region, and that yield gains have occurred without increasing the vulnerability of agro-ecosystems, especially due to their dependence on inputs (regularity and effectiveness of active ingredients). Moreover, the yield increases foreseen in Agrimonde GO could intensify pressure on water resources, possibly leading to the emergence of water-related conflicts between the domestic and agricultural sectors, crystallised around the domestic supply of drinking water to towns.

---

[1] According to the Mediterra Report, the urban population of the Southern Mediterranean increased from 108 million inhabitants in 1990 to 140 million inhabitants in 2000. It is estimated at 214 million inhabitants in 2020 (CIHEAM 2008).

[2] In North Africa, the average distance between towns shrunk from 66 km to 21 km between 1950 and 1995 in the coastal area, and from 66 km to 32 km inland (CIHEAM 2008).

[3] In this respect, Mediterra proposes, in order to "safeguard the land": setting up a dynamic and exhaustive system to monitor land trends (area and quality of agricultural and arable land); strengthening the effectiveness of legal systems capable of clearly distinguishing agricultural land from building land; and promoting the contracting of rural leases (CIHEAM 2008).

In both cases, Agrimonde 1 and Agrimonde GO, the agricultural scenarios for this region are subject to strong constraints.

Agriculture and Food in Middle East—North Africa: What are the Drivers of Change?

The Agrimonde 1 scenario is characterised by three main challenges in this region: water management, the development of rural employment, and the reduction of inequalities in access to food.

In the Agrimonde 1 scenario, between 2000 and 2050 the agricultural sector has been faced with the challenge of improving both added value and crop yields—by at least 13 % compared to 2000—in conditions of water stress accentuated by the effects of climate change. The problem of water resources and modes of agricultural production in conditions of water stress have therefore been a research priority in this scenario. This has resulted in scientific and technological breakthroughs in irrigation techniques, genetic selection, and agricultural practices which, in this region as in many others, have focused strongly on an improvement of water infiltration and on crop rotation requiring less water. In parallel, water management policies in this scenario have had to be reinforced decisively. They have made it possible to rationalise the allocation of the resource for various purposes, and to improve its use.

In Michel Griffon's Doubly Green Revolution (DGR) scenario, crop yields have increased by 0.59 % per annum owing to ecological intensification techniques (Griffon 2006). This gain is situated between the low variant of Agrimonde 1 and that of Agrimonde GO. The author emphasises the need to improve the efficiency of irrigation and to prioritise techniques for preserving water in the soil and plant coverage, mainly with legume crops, as well as the combination of cereal crop farming and livestock farming (especially ley farming[4]). He considers it inevitable that irrigated areas will be devoted more and more to vegetable crops and possibly to the production of fodder for intensive breeding. For this region, the Mediterra foresight study (CIHEAM 2008) proposes complementary ways to optimise the management of water resources (Box 9.1). The possibility of virtual water trade mentioned in this report emphasises the fact that the optimisation of water management also entails reflection on food trade and on the choices of agricultural policies.

In the Agrimonde 1 scenario, between 2000 and 2050 the rural population has grown considerably and most of this population has found employment in rural areas. The rural exodus has therefore been contained. This implies that the region has been able to re-balance development dynamics between the coastal areas and inland, characterised in the early twenty-first century by an urban shoreline linked to the international market where wealth was concentrated, on the one hand, and poor, uncompetitive inland areas, on the other. To deal with the social, economic and environmental problems associated with urbanisation that was difficult to control, regional planning and development policies in many countries of this region have endeavoured

[4] Ley farming is a technique developed in Australia, based on superficial tillage and cereal/alfalfa crop rotation for better crop-livestock integration (Lahmar 2006).

to maintain or trigger economic growth in rural areas by supporting agricultural and agri-food activities, as well as services, especially tourism in certain areas. The re-localisation of food systems has been one of several ways to integrate small farmers into the local market and thus to develop a solvent demand in rural areas.

Box 9.1—Proposals to optimise the management of water resources in the Mediterra Report (CIHEAM 2008)

In terms of policies on water supply management, development opportunities lie in:

- desalinisation of sea water for countries that can afford it,
- progress in reuse of urban waste water for agricultural irrigation.

Policies on demand management should promote:

- water saving (especially through technical means such as precision irrigation and improvement of water conveyance systems),
- policy trade-offs between uses, according to the cost-benefit ratio and positive externalities,
- rate innovations (rate scales).

Finally, the new water policy must be accompanied by legal and institutional changes:

- amendments to current regulations,
- creation of a water management unit in the watershed, with the support of local consumers of water,
- trading in virtual water.

Finally, diet is the last major change assumed in MENA in Agrimonde 1. Between 2000 and 2050, this region's mean per capita food consumption has dropped from 3,339 kcal to 3,000 kcal. We can imagine that effective nutritional policies have been implemented to curb the steady increase in obesity[5] (aimed mainly at reducing consumption of saturated fats and sugars). The reduction of income inequalities, and thus access to food, has had to be a key objective of social and economic policies. They have been complementary to regional planning and development policies in so far as they have focused on improving living conditions in rural areas.

Sub-Saharan Africa: Reconciling Sustainability and Development

In the Agrimonde 1 scenario, the population of the sub-Saharan Africa (SSA) region has multiplied by 2.5 between 2000 and 2050, while the per capita calorie consumption has increased by 30%. In this respect, the situation in Agrimonde GO

[5] "In the Southern Mediterranean, food availabilities have risen steeply over the past forty years, with a mean gain of 800 kcal/cap/day" (CIHEAM 2008).

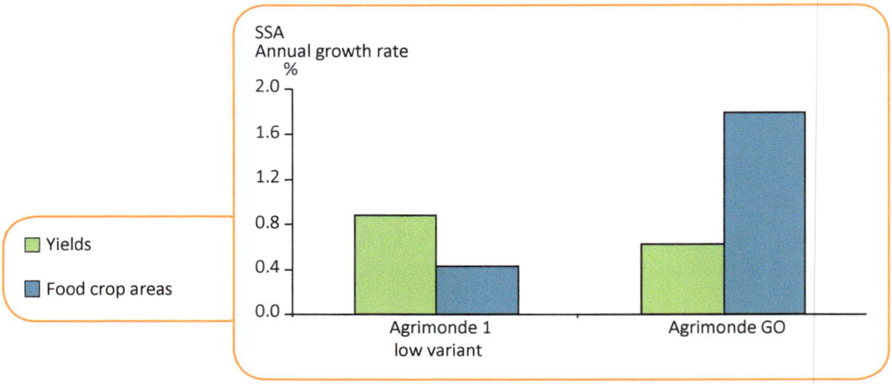

Fig. 9.2 SSA: Annual growth rate of food crop areas and yields in the Agrimonde scenarios between 2000 and 2050

is very similar, except that the proportion of animal and aquatic products in diet is lower (10% as opposed to 17% in Agrimonde 1). The total regional food demand in calories (see p. 172) is therefore virtually identical in the two scenarios (around 4,900 Gkcal/day).

In view of these sharp rises in needs, the Agrimonde 1 scenario foresees considerable expansion of agricultural areas rather than large food crop yield increases, while Agrimonde GO foresees a steep increase in yields and slightly less expansion of food croplands (roughly, 110 million additional hectares in 2050 in relation to 2000 in Agrimonde 1, compared to 70 million in Agrimonde GO).

The relatively low yield gains in Agrimonde 1 (+0.44% per annum in the low variant), especially compared to Agrimonde GO (+1.80% per annum, Fig. 9.2) and previous trends (+1.63% per annum between 1961 and 2000), stem from the following: first, the fact that potential impacts of climate change are taken into account, especially in West Africa, where certain studies foresee yield losses with no changes in techniques; and, second, the low capital intensity of the production systems considered in this scenario for this region. Similarly, in his DGR scenario, Michel Griffon foresees a yield gain (+0.62% per annum) equivalent to the low variant in Agrimonde 1. This leads him to assume a fivefold increase in cultivated areas, to achieve a regional balance of resources and uses in the case of plant calories for direct human consumption.

Total use of plant calories (see p. 172) in SSA rose from 2,000 Gkcal/day in 2000 to 7,500 in 2050 in the Agrimonde 1 scenario (+280%) and to 7,400 Gkcal/day in Agrimonde GO (+270%) (in Variant 1 of the resource-use balances of the scenarios, Chap. 8). On the whole, the strategies considered in the Agrimonde 1 and Agrimonde GO scenarios are not equivalent since the resource-use balance in 2050 is far more favourable in Agrimonde GO than in the low yield variant of Agrimonde 1: there is a shortage of some 1,300 Gkcal/day in Agrimonde GO, compared to almost 4,000 in Agrimonde 1.

Are the Agrimonde Scenarios Coherent in Sub-Saharan Africa?

The respective resource-use balances of these two scenarios beg the question of the coherence of the Agrimonde 1 scenario. Is it really a sustainable development scenario? Two main problems emerge in this respect.

First, in a context of high population growth but low yield gains, economic growth in the agricultural sector may not be enough to provide a basis for sufficient economic growth to ensure a solvent food demand. In particular, some agricultural households may not benefit from the development foreseen in Agrimonde 1. In that case, they would be marginalised[6], and would therefore be heavily affected by poverty and problems of access to food.

Second, even though the choice of moderate yield gains coupled with low-intensity production techniques in terms of mechanisation and inputs makes it possible to consider a scenario in which agriculture's impact on cultivated ecosystems is moderate, the large-scale conversion of pastures and especially of forests into cultivated lands probably has a considerable impact on the ecosystem services provided by these natural and semi-natural areas. The maintenance of soil fertility in the Agrimonde 1 scenario is particularly important, especially in areas taken from forests, where the land is particularly fragile due to the interruption of existing cycles of macro- and micro-nutrients.

In this respect Agrimonde GO, which foresees a distinct acceleration in the rate of yield gains between 2000 and 2050 compared to the period 1961–2000, assumes large-scale extension of farmland, although to a lesser extent than in Agrimonde 1. This conversion concerns mainly forest areas since the increase in animal product consumption in this region is reflected in an increase in pasture areas. Deforestation is therefore far more marked in Agrimonde GO (−200 million ha) than in Agrimonde 1 (−57 million ha). Thus, even an assumption of very high yields does not really make it possible to avoid a dwindling of forest areas in SSA.

Agriculture and Food in Sub-Saharan Africa: What are the Drivers of Change?

The problems raised by the coherence test on the Agrimonde 1 scenario are also the two main challenges in this region. Agrimonde 1 foresees two major changing trends between 2000 and 2050 compared to the period 1961–2000:

- agricultural development, which allows for a 30 % increase in mean per capita calorie consumption between 2000 and 2050,

[6] The 2005 UNDP report on human development shows that SSA is the region of the world with the most inequality. Its GINI index is 72.2 (UNDP, 2005). This coefficient measures the degree of inequality of income distribution in a given society. It ranges from 0 (perfect equality) to 100 (total inequality: one person concentrates all the income and the others have nothing).

- the accelerated conversion of pastures into croplands, while the conversion of forests has slowed down.

Development has clearly been the first challenge in SSA in the Agrimonde 1 scenario, as in the Agrimonde GO scenario. It begs questions that have already been raised countless times, on access to capital, techniques, land, training and markets, on the improvement of infrastructure and governance trends in this region, or on regulations of global trade in agriculture. The Agrimonde 1 scenario and its comparison with the Agrimonde GO scenario enable us to clarify some of these questions.

Agrimonde 1 is inspired by the DGR scenario put forward by Michel Griffon, in which production systems in SSA are low capital-intensive. Unlike the Agrimonde GO scenario, which promotes intensification systems based on large-scale mechanisation and use of inputs, the Agrimonde 1 scenario explores the development of technical systems that limit:

- the need for capital,
- dependence on fossil fuels that are expected to become scarce and therefore expensive,
- the impact of agriculture on ecosystems.

According to Michel Griffon, the main technical drivers of the DGR in this region are the creation of small water structures in dry areas, greater complementarity of crop and breeding systems in grassland areas, and the implementation of agro-forestry practices based on the "creole garden"[7] model in forest areas. Yet, according to this author, yield gains are more difficult to achieve on the highlands, which are currently experiencing an acute crisis in their agrarian systems[8]. He considers that technical systems will be able to improve yields by 0.62 % per annum between 2000 and 2050. This is close to the low yield variant envisaged in Agrimonde 1 (+0.44 % per annum). It is a huge challenge, especially since the crop systems will have to enable farmers to cope with the lack of water in certain regions and with the accentuation of climate variability.

As noted above, the yield gains foreseen in the low variant of Agrimonde 1 may however not be enough to allow for sufficient development. For Agrimonde 1 to be a scenario of development through agriculture in SSA, it seems that the high yield variant should be considered (+1.40 % per annum). However even if this variant is not particularly demanding compared to the Agrimonde GO scenario (+1.80 % per annum) or even past trends (+1.63 % per annum between 1961 and 2000), it is far higher than that considered by Michel Griffon. It raises the question of the possibility of developing ecological intensification technologies with which yields could be doubled in SSA by 2050.

Moreover, the technological systems of the Doubly Green Revolution are only partially available today, whereas classical intensification systems are available and, as in Agrimonde GO, are able to provide much higher yield gains than in

[7] The creole garden concept refers to a wide diversity of plants cultivated, often together, on a small surface area.

[8] This crisis of agrarian systems results in: small area per family, difficulty maintaining soil fertility, production caps, competition for use of resources (Griffon 2006).

Agrimonde 1, at least at the beginning of the period. Consequently, if agricultural development is the priority in SSA, should a two-phase intensification not be considered in this region? The first phase would consist in the diffusion of classic intensification techniques, which would enable agriculture to "take-off". It would be followed by a second ecological intensification phase once the techniques have been developed and the environmental challenges have become greater than those of development. The choice of such a trajectory raises the question of the obstacles to the green revolution, as well as that of the irreversibility of technological choices. The deployment of a technical system creates strong interdependencies between technologies, infrastructures, training, configurations of actors, and so on. This can result in lock-in that prevents or hinders technological changes even when the market environment and the potentialities of new technologies make the 'switchover' optimal from an economic point of view.

The changing trend in the cultivation of new lands is the second challenge of the Agrimonde 1 scenario in this region. With the objective of maintaining forest resources as far as possible, we assumed that by 2050 only 40 % of new land cultivated would be taken from forests and the rest from pastures (or grasslands). In addition to the 90 million ha of savannah converted to agricultural land in this scenario, the uses of savannah areas have become more complex due to large-scale agro-forestry practices, which have been an important element for production, and for employment and rural income. However, the type of organisation and ownership of land in place in 2000 made it difficult to stop the advancement of pioneering fronts on the forests of certain areas in West Africa (Southern Mali, Western Burkina, Eastern Senegal, Nigeria, Togo and Northern Ivory Coast). Pressure on forests has been especially intense due to their water resources, and because the impacts of climate change have shifted the cultivation potential away from pastured savannah towards these areas. The Agrimonde 1 scenario is therefore based on the creation of ecological infrastructures for the equatorial forest, which has allowed for the cultivation of new land and to some extent the conservation of biodiversity. Incentives to preserve the forest, which are therefore a key point in the scenario for this region, and for others, imply strong world governance.

Latin America: Resources to Exploit but Also to Protect and to Share

From 2000 to 2050, the population of Latin America (LAM) has grown by 50 % in both scenarios. In parallel, the total per capita food demand in calories has stabilised in Agrimonde 1 (−3 % compared to 2000) and increased in Agrimonde GO (+20 %). The total calorie demand (see p. 172) in 2050 is 2,300 Gkcal/day in the former and 2,900 Gkcal/day in the latter. Diets in these two scenarios differ from the point of view not only of total calorie demand but also of their composition. In Agrimonde GO one quarter of the calories are of animal and aquatic origin, compared to one sixth in Agrimonde 1.

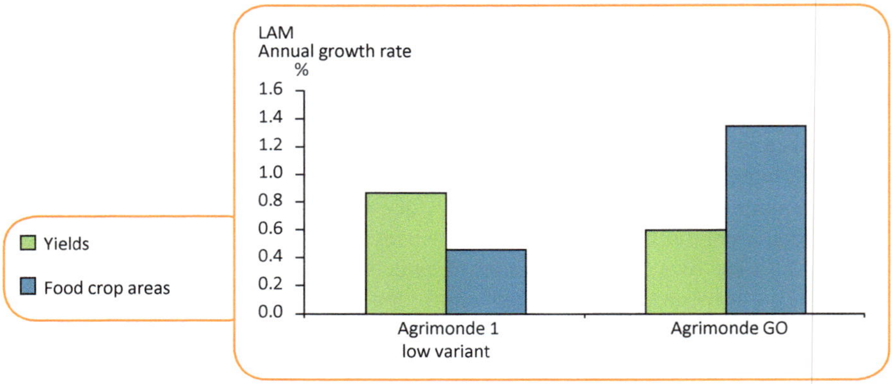

Fig. 9.3 LAM: Annual growth rate of food crop areas and yields in the Agrimonde scenarios between 2000 and 2050

In order to increase the region's productive capacities, the strategy adopted in the Agrimonde 1 scenario is based on the exploitation of part of LAM's enormous cultivation potential and on moderate food crop yield gains, while Agrimonde GO has essentially projected an increase in yield gains owing to technological progress. Cultivated areas have almost doubled in Agrimonde 1 (with an increase of only 54 % of land for food production), whereas they have increased by no more than 64 % in Agrimonde GO (with +36 % of the land for food production). Crop yields have increased by 26 % in Agrimonde 1 but almost doubled in Agrimonde GO.

Yield gains have slowed down compared to past trends in both scenarios (+0.46 % per annum in the low variant of Agrimonde 1 and +1.35 % in Agrimonde GO between 2000 and 2050, compared to +1.88 % per annum between 1961 and 2000, Fig. 9.3). In Agrimonde 1 this slowdown is imputable to the fact that the fragility of soils is taken into account, especially in grassland areas and on the forest pioneering fronts of this region. Progress is also limited by the effects of climate change which, according to the FAO, could cause the desertification and salinisation of 50 % of the land by 2050, with resulting yield losses (FAO 2004).

Total use of plant calories (see p. 172) in LAM rose from 2,700 Gkcal/day in 2000 to 4,000 in 2050 in the Agrimonde 1 scenario (+45 %) and to 5,900 Gkcal/day in

Agrimonde GO (+120 %) (in Variant 1 of the resource-use balances of the scenarios, Chap. 8). Overall, the two strategies adopted in the Agrimonde 1 and Agrimonde GO scenarios show an equilibrium in the resource-use balance in the region in 2050, and even comparable regional surpluses (of about 2,000 Gkcal/day).

Are the Agrimonde Scenarios Coherent in Latin America?

Is the doubling of cultivated areas in Agrimonde 1 compatible with sustainable development?

This question is raised regarding environmental issues, since it has resulted in a substantial reduction of natural and semi-natural areas, and also regarding social issues, as territorial extension could exacerbate glaring inequalities in access to land.

In the Agrimonde 1 scenario some 150 million ha of new land have been cultivated between 2000 and 2050. 110 million ha have been taken from former pastures, where the soil is often fragile (overgrazing), and the rest from forests. These transformations of natural and semi-natural areas are likely to impact on the biodiversity of ecosystems, on water and carbon cycles, and on the living conditions of communities dependent on the natural resources of these ecosystems (food, wood, pharmacopoeia, cultural symbols).

From a social point of view, in one of the regions with the most inequality in the world[9], many questions can be raised on the conditions of access to these newly cultivated areas and on the distribution of the rents derived from their exploitation. Have these new lands enabled landless families to set up farms? Have they been the locus of growth of food production and processing systems to ensure their inhabitants' food security? Or have they been put to the service of the development of biofuels (which have gained 60 million ha in the Agrimonde 1 scenario) and their industries? What has the role of the state been in regulating the rural land market?

These questions may seem less relevant in Agrimonde GO where pressure on uncultivated land is not nearly as intense as in Agrimonde 1 (− 13 million ha), but they do appear to be more relevant in the case of cultivated lands. Were the assumed yield gains in this scenario achieved without increasing environmental impacts (especially pollution through inputs)? To intensify production and possibly avoid its negative effects on the environment, have the technical systems implemented not been exclusively within the reach of very large concentrated farms? In this case, what is the social situation of small farmers, especially those who were landless in 2000?

Food and Agriculture in Latin America: What are the Drivers of Change?

In Agrimonde 1 questions of environmental and social sustainability are also the main challenges in LAM. Management of natural resources and land are therefore the main drivers in the implementation of the scenario.

In this scenario, we can imagine continuous encroachment on forests (especially the Amazon) initially, during an early phase in which the ecosystem services provided by the world's largest forest declined substantially. Apart from its function of storing carbon, the Amazon Basin's water-related functions were altered. The internal cycle of evaporation and rain was affected, and reduced the overall quantity of water circulating in the Amazon system, thus accentuating aridification phenomena

[9] The 2005 UNDP human development report shows that LAM is one of the regions with the most inequality in the world, second only to SSA. Its GINI index is 57.1, compared to 72.2 for SSA (UNDP 2005).

induced by climate change (Box 7.3). In response to these trends, strong global environmental regulations were implemented from 2015. Coupled with pressure from national and international ecologist lobbies, they precipitated the implementation of public policies for the protection, regeneration and management of forests. These policies aimed at putting a stop to the 'domino effect' triggered by the development of biofuels, i.e. the migration of crop farming supplanted by biofuel crop farming towards the forest pioneering front. They also aimed at controlling deforestation dynamics due to livestock farming, fires, the timber industry, and so on. Moreover, the maintenance of biodiversity implied that special attention be paid to the biodiversity hotspot in the Brazilian *Cerrado*, and that ecological corridors and regulation of the commercial exploitation of natural resources (wood, minerals, species with a pharmaceutical or cosmetic value, etc.) be set up.

The cultivation of tropical or newly deforested land also raises the question of how their fertility is reproduced, i.e. the fertilisation techniques adopted. Michel Griffon believes that in this region the techniques of the Doubly Green Revolution will entail the management of forest areas in which the forest's ecological, climatic and productive functions are reconciled. It will also involve improved mulching techniques, greater complementarity of cropping and breeding in traditionally agricultural areas in tropical and temperate climates, small-scale water structures, and integrated water management in dry areas (Griffon 2006). Thus, in the crop systems of the Agrimonde 1 scenario, expected agricultural innovations, rather than resulting in yield gains, result in a greater complexity of production systems on the frontiers of traditional farming areas: agro-pastoralism and farming systems in shaded areas of woodland savannah, and agro-forestry on the pioneer forest front. On the other hand, grazing-animal breeding systems have been intensified in Agrimonde 1 as the total demand for calories of grazing-animal origin has remained stable in LAM in the period between 2000 and 2050, with a 20% decrease in pasture areas. In view of these changes in crop and breeding systems, agronomic research has geared its studies towards production in forest areas and grasslands, and furthered its examination of specific complementarities and allelopathy in time and space. Environmental issues have also been a priority since the objective in this region has been to preserve forests and to limit the environmental impacts of intensified livestock farming on pastures which are shrinking in size and increasingly fragile due to climate change. Management of natural resources has evolved towards a greater awareness of the ecological value of natural resources, as well as their cultural content and the systems of belonging governing them.

The question of land tenure is closely linked to that of environmental sustainability. The trend towards monoculture over large areas is totally disrupting the functioning of ecosystems and impeding the maintenance of biodiversity. Moreover, the social sustainability of the Agrimonde 1 scenario requires a reduction of land inequalities. It implies that at least farmers' land security will be guaranteed, and that access to the rural land market will not be biased according to gender or ethnic origin. The efficiency of land use requires that access to capital be facilitated for those who have the fewest guarantees, and that stern efforts be made regarding agricultural training and the understanding of the functioning of supply chains and agricultural markets by all concerned.

Finally, for the social sustainability of the scenario, the question of the development of rural areas is a major issue in this region. The Agrimonde 1 scenario implies that agricultural development in newly anthropised areas has spawned rural jobs and that it has been an instrument for rural development with social policies and regional planning. In this region, the question of the actors of agricultural and rural development is decisive and is left wide open by the quantification of the scenario.

Asia: at the Frontier of Agricultural Potential, Controlling Urbanisation and Nutrition Transition

Between 2000 and 2050 Asia (ASIA) has experienced a population increase of 38% in both scenarios, bringing the total up to over 4.4 billion people. The evolution of per capita food consumption has varied widely: in Agrimonde 1 it has increased by 7% and in Agrimonde GO by 32%, i.e. by nearly 1,000 kcal compared to 2000. In this region the per capita consumption of animal and aquatic products in 2050 is twice as high in Agrimonde GO as in Agrimonde 1. The total regional food demand in calories (see p. 172) is therefore far higher in Agrimonde GO (16,400 Gkcal/day) than in Agrimonde 1 (13,300 Gkcal/day).

Faced with this growth of needs, and considering the very limited extension potential in 2000, cultivated areas have increased moderately in both scenarios although more in Agrimonde 1 (+23% between 2000 and 2050) than in Agrimonde GO (+11% for the same period)[10]. On the other hand, whereas pastures have decreased in Agrimonde 1 (−9%), they have increased in Agrimonde GO (+30%) to satisfy the high demand for animal products. These additional pastures have been taken from forest areas which have shrunk by 11% in Agrimonde GO and by 10% in Agrimonde 1.

Whereas Agrimonde GO foresees a steep increase in food crop yields (+1.23% per annum), Agrimonde 1 projects stable yields in ASIA in its low variant (Fig. 9.4). There are two reasons for this: first, the integration of the impacts of climate change, such as the frequency of violent events or other environmental problems like salinisation of deltas with irrigated rice paddies; and, second, the limits of the green revolution in this region which since 2000 has had the highest yields in the world (even if certain areas with rainfed agriculture can still experience yield gains).

Total use of plant calories (see p. 172) in ASIA rose from 11,500 Gkcal/day in 2000 to 16,700 in 2050 in the Agrimonde 1 scenario (+45%) and to 23,000 Gkcal/day in Agrimonde GO (+100%) (in Variant 1 of the resource-use balances of the scenarios, Chap. 8). In 2050 regional production fails to meet needs, in both Agrimonde 1 (approx. −3,000 Gkcal/day) and Agrimonde GO (approx. −900 Gkcal/day). ASIA has to import food to meet its animal product needs, in both scenarios. Imports are however three times greater in the Agrimonde 1 scenario than in Agrimonde GO, despite a far poorer diet in terms of calories.

[10] Strictly food-producing areas have grown by only 19% in Agrimonde 1 and 5% in Agrimonde GO.

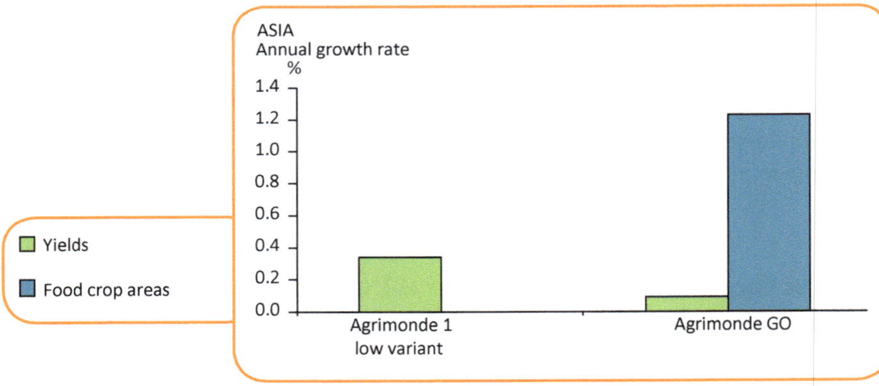

Fig. 9.4 ASIA: Annual growth rate of food crop areas and yields in the Agrimonde scenarios between 2000 and 2050

Are the Agrimonde Scenarios Coherent in Asia?

How coherent is the Agrimonde 1 scenario and what do we learn from the comparison between the two scenarios? Is the Agrimonde 1 scenario sustainable in ASIA? Between 2000 and 2050, cultivated areas have increased very little. As the agricultural surface area per worker was already very low at the beginning of the twenty-first century and yields were stagnant, the conditions seem conducive to a massive rural exodus consequent to population growth. This calls into question the coherence of our scenario in at least two respects:

- If agriculture absorbs only part of the Asian population growth, and people are unable to find occupations in rural areas, would a massive population shift to the urban areas be compatible with sustainable development?
- Urbanisation trends, which are expected to intensify if only due to the growth of the total population, are greater in Agrimonde 1 due to the principle of moderate migration in this scenario. In this light, is our assumption of an increase in cultivated areas realistic, since artificialised land will to a large extent be taken from cultivable land?

These questions on the rural exodus and urbanisation are equally relevant in the Agrimonde GO scenario and are considered by MA experts to be one of the major risks associated with this scenario.

Food and agriculture in Asia: What are the Drivers of Change?

Three main challenges characterise the Agrimonde 1 scenario in ASIA:

- the diffusion of environment-friendly agricultural practices while maintaining the yield levels,
- regional planning to contain urbanisation and the artificialisation of land,
- control of nutrition transition so that it does not result in the spreading of obesity.

Michel Griffon's yield assumption for ASIA in the DGR scenario is fairly optimistic since he foresees an increase of 0.79% per annum, which is close to the high variant of Agrimonde 1. He considers it possible to increase yields in humid tropical areas through forms of traditional agro-forestry and ecological management against runoff on slopes and on deforested pioneer fronts. In areas of rainfed agriculture he foresees the harnessing, conservation and re-accumulation of water in ecosystems, as well as micro-dams used for irrigation. For the tropical regions of the green revolution he cites the example of new techniques developed by the CIMMYT (International Maize and Wheat Improvement Centre), based on direct sowing and reuse of straws for the following crop[11]. We were not as optimistic as regards the potential of ecological intensification technologies. We considered that it would even be very difficult to maintain yields at the average level of 2000, with strict requirements for ecosystem preservation. Not only has the green revolution apparently already reached its limits in this region, but in addition climate change will aggravate salinisation, the lack of water which is already being felt acutely in the north-east of the region, and the risks associated with extreme climatic events (droughts, typhoons, etc.) whose effects are already accentuated by the on-going deforestation, especially in South-East Asia.

The effects of climate change on agriculture will have a decisive impact in ASIA since they threaten already fragile food security. In addition to the struggle against climate change and the adaptation of crop systems, in the Agrimonde 1 scenario this region has set up systems of stock regulation and trade at a regional level, as well as other measures to secure food imports. The implementation of such measures has certainly necessitated new regulatory mechanisms. For example, we can imagine that in the early twenty-first century, investments in farmlands were made by countries constrained by the need for cultivable space, in order to move towards areas whose potential was not yet completely exploited. This has implied the development of specific mechanisms of regulation and governance designed to guarantee the social and environmental sustainability of such investments, and to ensure that they become real development opportunities for host countries.

Agrimonde 1 describes a particularly fragile balance in ASIA, where agriculture does not seem to have absorbed the growth of the rural population, and the expected extension of cultivated land, albeit modest, has been difficult to implement due to the artificialisation of land. This scenario is therefore based largely on regional planning policies which have had two main objectives: the development of rural employment to curb the rural exodus; and the control of urbanisation, whether in terms of land competition with agriculture, management of water resources, problems of congestion, or social tensions. This calls into question the sustainability of urban and peri-urban forms of agriculture, the urbanisation models (vertical densification of megalopolises, development of medium-sized towns throughout the country, etc.) and, more generally, different configurations of city-country relationships likely to emerge.

The third challenge highlighted by the Agrimonde 1 scenario and its comparison with Agrimonde GO concerns nutrition transition. In Agrimonde 1 the mean

[11] Techniques developed on 6 million ha in 2004 (rice/wheat) or (rice/maize) (Griffon 2006).

demand for calories in 2050 in all the world's regions is around 3,000 kcal/cap/day. Agrimonde GO, on the other hand, foresees a trend-based scenario as regards diet. Hence, whereas in Agrimonde 1 the per capita calorie demand in ASIA has increased by 7 %, in Agrimonde GO it has risen by 32 %, driven by rising incomes and the generalisation of an urban lifestyle, and consequently of the growing proportion of animal calories in diets. The question is therefore: what factors could curb the increase in food consumption? This question is relevant to all developing regions but ASIA in particular, whose population accounted for over half the world population in 2000 and is expected to increase by almost 40% between 2000 and 2050. It is therefore in this region that the difference in plant food calorie use (see p. 172) in 2050 between the two scenarios is greatest (respectively, 16,700 and 23,000 Gkcal per day). The possibility of containing the increase in the mean food consumption in ASIA (especially the share of animal calories) is decisive in two respects in the Agrimonde 1 scenario: the world's food resource-use balance depends heavily on it; and sustainability from the point of view of health is likewise dependent on it, if we consider that the spreading of obesity resulting from the early twenty-first century nutrition transition in developing countries could get worse. The rate of adults who became obese every year in China between 1992 and 2007, for example, was higher than that of both developed and developing countries, except for Mexico (Popkins 2008).

Former Soviet Union: A Reservoir of Production but an Agricultural Model to Adapt to a Declining Population

Between 2000 and 2050 in both scenarios the population of the former Soviet Union (FSU) has dropped by 15%. In Agrimonde 1 this region has maintained its per capita food calorie demand, with a total demand (see p. 172) of up to 700 Gkcal/day. In Agrimonde GO the per capita food calorie demand has increased by 13% compared to 2000, and total demand attains 800 Gkcal/day in 2050. Moreover, diets comprise a larger proportion of animal and aquatic calories than in the Agrimonde 1 scenario (respectively, 40 and 17% of calories of animal and aquatic origin in Agrimonde GO and Agrimonde 1). The total animal and aquatic calorie demand has thus dropped by 30% in Agrimonde 1 but climbed by 87% in Agrimonde GO.

In the Agrimonde 1 scenario, FSU has put a stop to the reduction of cultivated areas and food crop yields observed between 1961 and 2000. Its agriculture has really taken off: cultivated areas have increased by 53 % and yields have doubled between 2000 and 2050 (+1.33 % per annum), resulting in the conversion of slightly more than 100 million ha. This is based on two assumptions: that pasture areas have moved northwards as the permafrost melts, and that the region has caught up with its 1990 yield level.

In contrast, in the Agrimonde GO scenario, the growth of production is the result of yield gains only (+1.09 % per annum between 2000 and 2050), even though these

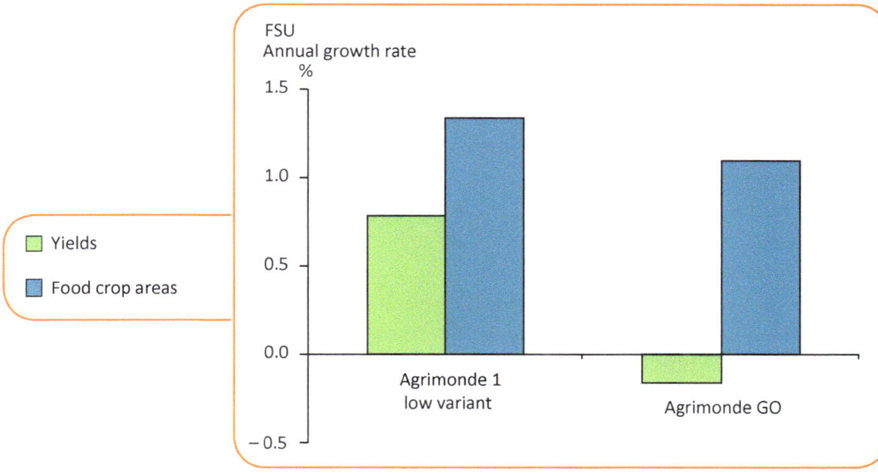

Fig. 9.5 FSU: Annual growth rate of food crop areas and yields in the Agrimonde scenarios between 2000 and 2050

gains are not as high as those foreseen in Agrimonde 1 (Fig. 9.5). The logic of Agrimonde GO is very different, since the conversion of new land aims to balance the supply and demand of biomass, and not to take advantage of the region's huge cultivation potential. Cultivated areas have consequently grown by only 20 million ha between 2000 and 2050. Thirty-six million hectares are devoted to biofuels in 2050, thus reducing foodcrop areas by an equivalent surface.

In both scenarios the extension of pastures observed between 1961 and 2000 has ended. In Agrimonde 1 new grasslands in areas where the permafrost has melted are probably less productive than former pastures converted to croplands, but the reduction of pastures is associated with a 40% reduction in individual demand for calories from grazing animals between 2000 and 2050. In contrast, in Agrimonde GO this demand has more than doubled during the same period.

In parallel, the current deforestation trend has been interrupted between 2000 and 2050 in Agrimonde 1 and has been reversed in Agrimonde GO where forests have gained 100 million ha. Boreal forests have thus been preserved in both scenarios. These forests "constitute the main carbon reservoir in the biosphere. They are distinguished from temperate and tropical forests by the fact that they store huge amounts of carbon in the soil (85% compared to 30–75% for the others)" (D4E 2006).

Total use of plant calories (see p. 172) in FSU has declined from 1,500 Gkcal/day in 2000 to 1,000 in 2050 in the Agrimonde 1 scenario (−30%) and increased to 2,100 Gkcal/day in Agrimonde GO (+45%) (in Variant 1 of the resource-use balances of the scenarios, Chap. 8). In both scenarios uses of biomass can easily be covered by regional resources in 2050. The surpluses are almost 12 times greater in Agrimonde 1 (3,300 Gkcal/day) than in Agrimonde GO (300 Gkcal/day).

Are the Agrimonde Scenarios Coherent in the Former Soviet Union?

Comparison of the two scenarios begs the question of the significance of the conversion of over 100 million ha (in Agrimonde 1) in a region in which the population is decreasing. Two possibilities can be considered: the cultivated area per farm worker could increase steeply as a result of automation and mechanisation of agricultural tasks; and farmers may employ large numbers of labourers from the border areas of Asia and Europe.

These two options raise questions on the coherence of the Agrimonde 1 scenario. The adoption of new techniques which can be substituted for certain human tasks and make it possible to expand the cultivated areas per farm worker requires large capital investments. This may not be possible for all farmers and could therefore lead to a two-speed development: while some farmers expand and equip their farms, others gradually decapitalise to remain competitive. Moreover, in this context of new productive spaces emerging with a decreasing population, the rural fabric could be eroded to a large extent, leading to all sorts of problems of access to services, isolation of people, etc.

The option of using labour from other regions goes against the assumption of an absence of major inter-regional migration on which the Agrimonde 1 scenario was based (Chap. 4). This option could however be a solution to the problems of agricultural employment emerging in ASIA due to the limitations of the cultivation potential. Labour employed in FSU could logically be of Asian origin in this scenario. In this case, we would however be departing from the initial logic of assessing "the effects of future demographic trends, without them being masked by large international migratory flows" that we chose to explore. In any case, this region's role as a reservoir of production in the Agrimonde 1 scenario is a particularly strong assumption.

Agriculture and Food in the Former Soviet Union: What are the Drivers of Change?

As we have seen, the main challenge highlighted by Agrimonde 1 in this region is the cultivation of over 100 million ha of converted land in 50 years, in a context of population decline. In this situation the doubling of yields is probably a result of highly automated agriculture using imagery and information technologies to replace the observation and analysis of the state of crops, and the development of precision techniques in agriculture to limit chemical pollution.

In Agrimonde 1, the agricultural landscape has tended to be simplified, with more cultivated fields in the south of the region and more pastures in the north. This redistribution of agricultural activities has probably been attended by a specialisation of farms. In plant production, farmers' specialisation has been compatible with the expected extension of the surface area farmed per farm worker (increased productivity of labour). In livestock farming it may have resulted in more extensive ruminant breeding systems per hectare on less productive land where the permafrost has melted. Between 2000 and 2050, the total demand for calories from grazing animals has indeed been halved, and pastures have been reduced by 16%. This evolution has required agronomic research to meet the challenge of minimising the negative externalities of an agriculture which has at least doubled crop yields by developing specialised production systems and even by separating crop farming from livestock farming. The farming of fragile land freed by the melting of the permafrost has furthermore created new challenges concerning soil conservation.

To grasp the opportunities of territorial extension in the north resulting from climate change, adaptations have had to be made, particularly:

- a legal framework and land policy have had to be implemented to govern the redistribution of new land formerly under permafrost,
- the network of transport infrastructure has had to be extended to new productive areas. This may have included the construction of new ports on the currently frozen Arctic coast.

OECD-1990: Diet, an Important Element in the Choice of a Development Model

Between 2000 and 2050, in both scenarios, OECD-1990 (OECD) has experienced a 10% population growth, i.e. 94 million additional inhabitants. The per capita consumption of food calories has declined by 24% in Agrimonde 1, to 3,000 kcal/cap/day, with a 60% reduction in the consumption of animal and aquatic products. In contrast, in Agrimonde GO the consumption of food calories has remained sTable (+4%) and that of animal and aquatic products has increased by 40%. In 2050 the total food demand in calories (see p. 172) is 4,400 Gkcal/day in Agrimonde GO and 3,200 Gkcal/day in Agrimonde 1.

The evolution of land use is similar in the two scenarios which foresee a decrease in both food croplands (approximately 5% in both scenarios) and pastures (approximately 20%), while forests have grown (about 10% in both scenarios). Food crop yields have evolved to a different extent in the two scenarios: stability in the low variant of Agrimonde 1 and a steep increase in Agrimonde GO (+48%, i.e. +0.79% per annum, Fig. 9.6).

In the low yield variant of Agrimonde 1, the stabilisation of yields in OECD has primarily been the result of the priority given to quality products from organoleptic, sanitary and environmental points of view, rather than to the intensification of production systems.

Total use of plant calories (see p. 172) in OECD declined from 8,500 Gkcal/day in 2000 to 4,900 in 2050 in the Agrimonde 1 scenario (−45%) and increased to 10,900 Gkcal/day in Agrimonde GO (+30%) (in Variant 1 of the resource-use balances of the scenarios, Chap. 8). Despite a steep increase in yields, the total surplus

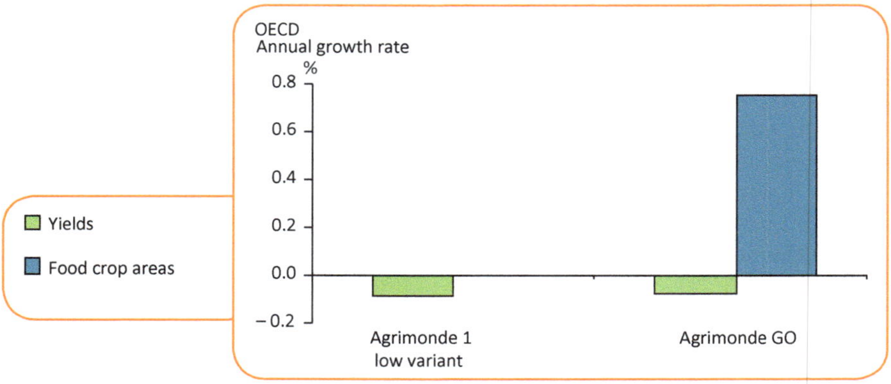

Fig. 9.6 OECD: Annual growth rate of food crop areas and yields in the Agrimonde scenarios between 2000 and 2050

in 2050 in Agrimonde GO is nearly half that of Agrimonde 1 (4,200 Gkcal/day in Agrimonde 1 and 2,500 Gkcal/day in Agrimonde GO). The reason is the difference in the consumption of animal products in the two scenarios. Thus, in the OECD region, the comparison of the two scenarios most clearly highlights the impact of the consumption of animal products on food resource-use balances, particularly in view of the importance of animal feed in livestock farming in this region.

Are the Agrimonde Scenarios Coherent in OECD-1990?

The comparison of production potential with consumption in the Agrimonde 1 scenario reveals a reduction in calorie consumption per inhabitant, whereas the region is far from attaining its production limits in terms of surface area as well as yields. While this poses no problem of coherence on a regional scale, we must question the coherence of such a regional option on a global scale. Stronger pressure on resources in the OECD region may have eased pressure on ecosystems in other regions.

Agriculture and Food in OECD-1990: What are the Drivers of Change?

Two main challenges characterise the OECD region in the Agrimonde 1 scenario: a change of direction for agriculture as regards both its practices and the products and services offered; and the widespread adoption of lower-calorie diets, with particular reduction of calories of animal origin.

The assumption of stable yields through changes in agricultural practices, with less intensive use of chemical inputs and based more on knowledge of ecosystems, raises questions on the practices implemented, on the ability of these practices to meet the objective of maintaining yield levels at their 2000 level, and on the drivers of such changes. This evolution can be explained by consumer pressure and public policies to drive agriculture towards multifunctionality, less environmentally harmful practices, and the production of quality food, even if some farms have continued their industrial dynamics and their production of standardised food. This orientation nevertheless implies that products with a high added value have found enough outlets. While the considerable expansion of supply chains offering agricultural products with a high added value has probably allowed for economies of scale and thus a lower retail price, the Agrimonde 1 scenario also assumes that the growing precariousness of the middle classes in many countries of this region since the late twentieth century has been curbed.

The reduction by a quarter of mean per capita calorie consumption between 2000 and 2050 foreseen in Agrimonde 1 is of course a departure from the main trend in this region. It could be a result of nutritional policies even if their effectiveness is currently highly controversial, and of the reduction in consumption losses through less waste and better recycling of waste.

Increasingly Strong Regional Interactions to Feed the World

In both scenarios the world's population has grown by 50% between 2000 and 2050, attaining nearly 9 billion inhabitants, whereas the per capita demand for food calories varies from stagnation at 3,000 kcal/cap/day in Agrimonde 1 to growth to 3,588 kcal/cap/day in Agrimonde GO, i.e. a 20% increase compared to 2000. In 2050 diets in these two scenarios differ as regards not only the total calorie demand, but also their composition: 500 kcal/cap/day of products of animal and aquatic origin in Agrimonde 1 and 890 kcal/cap/day in Agrimonde GO. Total food calorie consumption (see p. 172) at global level is 26,400 Gkcal/day in Agrimonde 1 and 31,600 Gkcal/day in Agrimonde GO.

In the Agrimonde 1 scenario, production increases have been based on the exploitation of the cultivation potential, still uncultivated in 2000 in certain areas of the world (+39% of newly cultivated areas between 2000 and 2050), and the implementation of production techniques that allow for moderate yield gains (+0.14% per annum, Fig. 9.7). A different strategy is proposed in Agrimonde GO, where the expansion of cultivated surfaces has been more limited (+23%) and yield gains have been much higher (+1.14% per annum). Pasture areas have decreased by 15% in Agrimonde 1 but have increased by 7% in Agrimonde GO. These trends also reflect the different orientations taken in the two scenarios in terms of demand for calories from grazing animals (1,600 Gkcal/day in Agrimonde 1 and 4,100 Gkcal/day in Agrimonde GO). Forest areas have remained largely unchanged in both scenarios (−1%).

Total use of plant calories (see p. 172) in the world rose from 28,100 Gkcal/day in 2000 to 37,600 in 2050 in the Agrimonde 1 scenario (+35%) and to 53,600 Gkcal/

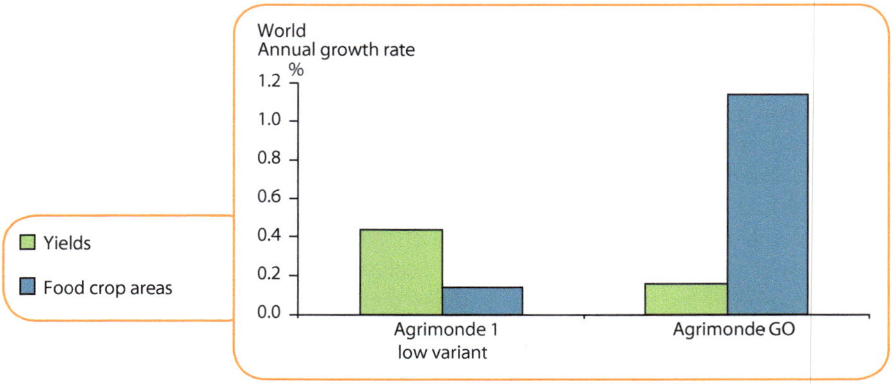

Fig. 9.7 World: Annual growth rate of food crop areas and yields in the Agrimonde scenarios between 2000 and 2050

day in Agrimonde GO (+90 %) (in Variant 1 of the resource-use balances of the scenarios, Chap. 8). In the final analysis, the two strategies allow for a global equilibrium in the food resource-use balance in 2050. In Agrimonde GO there is even a surplus (440 Gkcal/day). Inter-regional trade is however indispensable in both scenarios to meet the needs of all the regions.

Comparison of the two strategies clearly illustrates two possible modes of intensification. In the Agrimonde GO scenario, large yield gains make it possible to limit the conversion of natural areas (forests) and semi-natural areas (savannah or semi-natural pastures). Note however that 550 million ha of newly cultivated land come from the category 'other', i.e. they are a result of the conversion of steeply sloping or semi-desert areas not counted in the cultivated, forest or pasture areas in 2000. On the other hand, in the Agrimonde 1 scenario the wish to limit pressure on natural resources in productive areas has limited yield gains, especially in the case of the low yield variant (+7 % on average at a global level between 2000 and 2050). The two scenarios clearly show that to attain a food resource-use balance on a global scale, the extension of agricultural areas is indispensable, even if it is moderated by the stabilisation of the demand for food calories at around 3,000 kcal/cap/day in the Agrimonde 1 scenario, and by yield gains in Agrimonde GO.

Are the Agrimonde Scenarios Coherent on a Global Level?

In addition to the questions of coherence identified on a regional scale, the Agrimonde 1 scenario raises questions of global coherence:

- to be coherent, the Agrimonde 1 scenario is based on the possibility of inter-regional trade, as certain regions have a surplus while others cannot meet their food needs in 2050. These possibilities of trade are based on regulations which do not result in price distortions unfavourable to the development of agriculture

in developing countries. However, they provide for significant exceptions to enable the least productive countries to develop a local market, and make it possible to reveal the environmental costs of agricultural activities so that farmers are encouraged to develop sustainable agricultural systems,
- the disparity of levels of pressure exerted on the land in the different regions of the world calls into question the choice of no longer exploiting the cultivation potential of the regions LAM and OECD; this could in turn relieve the heavy pressure on land in other regions of the world (MENA and ASIA in particular),
- the scenario assumes that policies have been implemented to stop deforestation in FSU, to slow it down substantially in LAM, and to curb it slightly in SSA. This raises the question of the harmonisation of policies to manage natural resources on a global scale, and the articulation of regional and global initiatives,
- production of grazing animals increases in both scenarios even though the increase is far greater in Agrimonde GO. The reduction of pasture areas in Agrimonde 1 and their slight increase in Agrimonde GO may then seem contradictory since Agrimonde GO assumes an intensification of livestock farming. It raises questions on the most sustainable options for livestock farming. These should allow for optimisation in terms of the use of resources and the conversion of plant calories into animal calories while minimising environmental impacts (due to pollution and changes in land use). Moreover, such optimisation must be adapted to the local context. For example, in certain areas livestock farming has other functions, in addition to being a source of food for humans, such as animal traction, live capital, fertiliser, etc.,
- the strong contrast between regions in terms of population density also calls into question the assumption of unchanged migration levels up to 2050, on which the scenario was built.

The coherence of the Agrimonde GO scenario is based on two very strong assumptions. First, it assumes that trade liberalisation, coupled with significant technological progress, has boosted development. Second, the considerable yield gains foreseen on the basis of the continuation of the scientific and technological trajectory of the green revolution are sometimes contradictory to the projected yield losses due to climate change in the studies of the IPCC and of Parry et al. (2004), admittedly performed at a constant technological level (Boxes 7.2 to 7.7). This implies that, to be feasible, Agrimonde GO must always afford opportunities for considerable progress, and that the state of the ecosystems will not be a threat to the realisation of these opportunities.

What are the Drivers of Change for Agriculture and Food on a Global Scale?

On a global scale, the Agrimonde 1 scenario's main challenge is the inter-regional regulations and modes of governance that would enable it to emerge. These concern development aid, trade in agricultural goods, the struggle against climate change, the protection of ecosystems and natural resources (water and forests, mainly), land distribution and, finally, direct foreign investments in agriculture, including in its social and environmental dimensions. The question of intellectual property rights is also decisive as Agrimonde 1 is based on radical innovations which imply technical progress (and therefore incentives) and the diffusion of knowledge. Moreover, considering that food trade is indispensable, the Agrimonde 1 scenario carries the risk of trade becoming a food weapon. To mitigate this threat, international agreements on food policies must be signed.

Agrimonde GO foresees the liberalisation of trade and inter-regional transfers of development aid, combined with strong, multilateral world governance to accelerate growth in all the world's regions and thus reduce poverty and malnutrition. The Agrimonde 1 scenario is more demanding in terms of effectiveness of world governance which must also ensure pro-activity in the management of ecosystems, absent in Agrimonde GO. It is however difficult to take a clear stand on the multilateral or regional nature of governance in Agrimonde 1. Likewise, as Agrimonde 1 is based on the free circulation of agricultural goods (with exceptions to enable the least productive countries to develop a local market), the strong interactions between agricultural production and natural resources (and climate change) must be taken into account and bear upon the regulation of that trade.

Chapter 10
Qualitative Dimensions and Agrimonde Scenario Storylines

Sandrine Paillard and Sébastien Treyer

This chapter explores the variety of qualitative dimensions of the Agrimonde scenarios in order to propose complete scenarios at global level. First, on the basis of the exploration carried out in the preceding chapters, we examine the possible assumptions on the qualitative variables of the agricultural and food systems for each of the scenarios. Second, a possible storyline for each of these two scenarios is proposed.

The Qualitative Dimensions of the Global Scenarios

The dimensions and variables of the Agrimonde system presented in Chap. 1 (Table 1.1, p. 19) had to be analysed in depth to complete the quantitative structure of the scenarios with assumptions on qualitative trends. This analysis, carried out at global level only, for the sake of simplicity (and presented in detail in Appendix 6, p. 273), enabled us to construct the scenario storylines proposed in this chapter. Numerous variables are left undetermined in the quantitative scenarios and may be the subject of various assumptions for the future, provided that each scenario is coherent. It is then the principles on which the scenarios are built and the search for plausibility which guide the formulation of qualitative assumptions. Thus, alongside the quantitative scenario, the qualitative assumptions seem to determine whether the scenarios presented may represent possible futures or not.

This analysis enables us to emphasise certain salient points of the Agrimonde exploration. It highlights those assumptions for the future where the Agrimonde 1

S. Paillard (✉)
ANR, 212 rue de Bercy, 75012 Paris, France
e-mail: sandrine.paillard@agencerecherche.fr

S. Treyer
IDDRI, 27 rue Saint Guillaume, 75337 Paris Cedex 7, France
e-mail: sebastien.treyer@iddri.org

S. Paillard et al. (eds.), *Agrimonde – Scenarios and Challenges for Feeding the World in 2050*,
DOI 10.1007/978-94-017-8745-1_10,

and Agrimonde GO scenarios contrast the most, including the most marked changes envisaged in the former scenario by comparison with current trends that Agrimonde GO prolongs. Changes concerning the rural exodus, food-related behaviours, public policies and regulations, and innovation systems are some of the strongest assumptions in the Agrimonde 1 scenario:

- In the Agrimonde 1 scenario the world in 2050 has seen today's developing countries moving ahead, propelled by an agricultural progress characterised by essentially labour-intensive farming systems. Rural employment, apart from agricultural jobs, has also had to be a priority in development policies, especially in regions facing a limited agricultural growth potential, such as Asia (ASIA) and Middle East—North Africa (MENA). Thus, even if exodus and urbanisation in regions with steep demographic growth remain widespread, in Agrimonde 1 their pace tends to stabilise, while in Agrimonde GO they are accelerating as a result of very rapid technological progress in agriculture, geared towards capital-labour substitution.
- Agrimonde GO is intended to be a trend-based scenario as regards diet, since food consumption, especially food of animal origin, increases with income. Agrimonde 1, on the other hand, foresees major changes in diet linked to environmental and above all nutritional concerns, with the struggle against obesity being a key objective. While the quantitative assumptions concern only the mean number of kilocalories consumed in the different regions, and their distribution in terms of origin (plant, non-grazing animals, grazing animals, aquatic), it seems probable that the major changes marking them are also accompanied by major evolutions in food consumption practices and, more generally, in lifestyle, especially in regions marked by a decrease in total calorie consumption,
- As regards public action, Agrimonde GO is characterised by less intervention. Public action is generally reactive, whether it concerns nutrition, the environment, or energy. In Agrimonde 1, trade in agricultural products must be strongly encouraged since it needs to increase to ensure food security in all regions, compared not only to the beginning of the century but also to Agrimonde GO. Direct support for production is therefore destined to disappear (it was allowed only during the agricultural development phase in those countries most dependent on agriculture). The liberalisation of trade has not however been accompanied by less government intervention. Public intervention has been decisive and proactive, and aimed at regional development, ecosystem protection, and climate change adaptation and mitigation. Finally, nutrition-related policies are also highly ambitious.
- In these two scenarios the public- and private-sector research and innovation effort in the food and agricultural field has had to be massive and largely international. It has been complementary to heavy investments in training farmers in developing countries. The two scenarios contrast sharply however with regard to the nature of the knowledge produced and its modes of production and dissemination. In this respect the scenarios are consistent with the final report of the International Assessment of Agricultural Knowledge, Science and Technology for Development (IAASTD), which highlights the implications of different

types of innovation (IAASTD 2009). While in Agrimonde GO the objective of innovations is primarily to increase yields, in Agrimonde 1 these increases must be compatible with the objectives of protecting ecosystems and reducing dependence on inputs. Ecological intensification, with the decisive progress in knowledge on ecosystems that it implies, is based on a transformation of modes of knowledge production and dissemination characterised by strong synergies between local, lay and pluridisciplinary knowledge on the one hand, and specialised scientific knowledge on the other.

Our analysis of the qualitative dimensions of the scenarios also underscores the limits of the exploration carried out in the framework of Agrimonde. Some of these dimensions appear to be a determining factor in the plausibility of the scenarios, yet particularly difficult to apprehend, due to the macroscopic nature of the approach or to mechanisms pertaining to resource-use balances, which by nature are more a matter of accounting than of economics. Thus, for instance, the principles of scenario-building imply highly contrasting social forms of production in the two scenarios. While Agrimonde GO is characterised by continued industrialisation of agriculture, Agrimonde 1 assumes production techniques in developing countries which are not capital-intensive, and is based on more varied social forms of production, with a strong component of peasant agriculture in these countries. The modalities of coexistence of differing social forms nevertheless still need to be imagined, even though they are determining factors in the realisation of the scenario, especially in Latin America (LAM) and sub-Saharan Africa (SSA). Likewise, although the diversity of entrepreneurial forms in the agri-food industries is a key element of the Agrimonde 1 scenario, it adds little to the analysis of the coexistence and relations between the various industrial actors (SMEs, cooperatives, multinational firms), because our scenario-building focused above all on the production and use of biomass.

Finally, analysis of the qualitative dimensions of the scenarios enables us to question their sustainability. Sustainable development is a basic assumption in the construction of the Agrimonde 1 scenario, which starts with the idea that sustainable food and agriculture characterise the world in 2050. In both scenarios decisive progress has been made in alleviating poverty and malnutrition. In Agrimonde GO, this objective has been met to the detriment of other objectives such as the protection of ecosystems, the struggle against climate change, or bringing the obesity epidemic under control. In contrast, Agrimonde 1 explores the complementarity of these objectives, which is fragile: not only does it demand radical changes in behaviour, but the protection of ecosystems in certain areas runs the risk of limiting agricultural development, as seen in the previous chapter for SSA. Moreover, the choice of innovations that limit the impacts of agricultural activities on the environment has resulted in limited yield gains and consequently in a considerable expansion of cultivated areas. While this may have improved domestic biodiversity, it has certainly also weakened wild biodiversity. Moreover, changes in land use, characterised by a high level of conversion of pastures into croplands, have obviously resulted in a lower carbon storage capacity which has to be offset by production systems that rely less on fossil fuels and store carbon more effectively.

The World's Food and Agriculture in 2050 in the Agrimonde Scenarios

Analysis and comparison (Chap. 9) of regional quantitative scenarios in terms of coherence and drivers of change enabled us to refine certain qualitative assumptions in the scenarios. The analysis, reported above, on the dimensions and qualitative variables of the Agrimonde system, made it possible to move further towards the definition of plausible and coherent global scenarios. On this basis, a storyline of the Agrimonde GO scenario is proposed here—it is based on the Millennium Ecosystem Assessment (MA) experts' storyline of the Global Orchestration scenario (MA 2005b)—as well as a possible storyline for Agrimonde 1.

Agrimonde GO: Feeding the World by Making Global Economic Growth a Priority

In Agrimonde GO the world is preoccupied above all with the problem of employing and feeding a growing population. Human impacts on ecosystems increase as agricultural and urban surfaces spread.

Development and International Trade are Mutually Reinforcing

Since the early twenty-first century many have argued that hunger is a problem of equitable distribution rather than one of under-production. The strategy of increasing production (by increasing yields and, to a lesser extent, agricultural areas) has however weighed more in international debates and has been considered as the most effective way of alleviating poverty and achieving greater equity.

Huge investments in research and infrastructure, especially in developing countries, coupled with free trade have made it possible to meet sharp rises in food demand and to lower international prices of animal products and rice. The agricultural goods market has been totally liberalised, with the removal of many customs barriers and competition distortions. The benefits of trade liberalisation have rapidly become apparent and, as agricultural activities have developed and become more modernised, regions such as LAM and SSA have become net exporters of certain products, while OECD-1990 (OECD) and ASIA have seen their net imports increase.

Certain countries whose exports were based on tropical products (e.g. coffee, cocoa) have not benefited from international free trade. Economic growth in this group of countries is based on other opportunities, which have made it possible to diversify economic activities. In many cases these opportunities have however been limited due to a lack of policies in favour of human capital and infrastructures.

Economic growth has been very intense, surpassing previous averages in several regions (mainly SSA, ASIA and the former Soviet Union (FSU)). This growth is a result of the combination of trade liberalisation, extensive economic cooperation, and the rapid diffusion of new technologies. Investments in education and health have moreover been huge in all regions. ASIA has experienced a rapid growth rate (5–6 % annual growth of its GDP) throughout most of the first half of the century. LAM, which has overcome its problems of debt and trade deficits, has seen its per capita GDP rise by 3–4 % per annum, on average. SSA has implemented institutional reforms which allowed for a high growth rate after 2025, with the per capita GDP rising by 2–4 % per annum. FSU has relied on its qualified workforce to overcome its economic decline and to boost its economy (3–5 % annual growth of the per capita GDP between 2000 and 2050).

Very Rapid Technological Progress

Massive investments in education and research have resulted in the rapid creation of new technologies, and low trade barriers have facilitated their swift diffusion. Multinational firms are deeply involved in innovation. This goes hand-in-hand with powerful entrepreneurship dynamics but is not geared towards protection of the environment.

A Steep Increase in the Demand for Fossil Fuels and Biofuels

In this rapid growth scenario, a steep rise in energy demand has been experienced. This demand is mainly for fossil fuels but technological progress has resulted in lower average production costs for energy and has substantially improved energy efficiency. Electricity is partially produced from renewable sources (10 % of the total energy production in 2050) and biomass. Areas under biofuel crops have expanded considerably compared to the beginning of the century, driven by the rising prices of fossil fuels, while the intensification of food production has left land available for the production of biofuels, without further deforestation.

The Environment and Global Changes: Objectives that are Considered as Secondary

The conditions for environmental problems—especially those associated with climate change and fishing—to be addressed at global level are met through international cooperation. However since environmental problems are considered secondary to other priorities—economic growth, improvement of humans' material well-being—the problems threatening human welfare (pollution, erosion, climate change) are taken into consideration only when they become unavoidable. As international institutions have a reactive approach to ecosystem management, they are unprepared for unexpected ecological events. For instance, no climate policy exists in this scenario, and during the first decades no attempt is made to control greenhouse gas emissions. Trust in the ability of science and technology to meet environmental challenges creates a context favourable to ecological "surprises", such as the emergence of infectious diseases. Strong international cooperation therefore exists to control sanitary crises. In 2050 the increase of wealth has nevertheless finally resulted in greater concern among citizens with regard to environmental protection.

Continuation of Early-Century Trends in Diet and Consumption Habits

The global availability of food calories, per day and per capita, has increased by 20 % between 2000 and 2050, from 3,000 to 3,600 kcal. The steepest increases have been experienced in ASIA (+900 kcal/cap/day), LAM (+600 kcal/cap/day) and SSA (+650 kcal/cap/day), and the number of children suffering from malnutrition in developing countries was divided by 2.5 during the first half of the century.

This tendency, stimulated by rapid economic growth and intense urbanisation, is accompanied by a richer protein content of diets as people consume more meat and fish. Only in SSA and MENA has meat consumption not increased substantially. This trend has resulted in the growth of the problem of obesity in many regions (ASIA, SSA). Nutrition policies therefore need to be implemented to encourage physical activity and less consumption of fatty food.

Agricultural and Food Systems are Industrialised and Standardised

Technological development has allowed for more intensive farming and for extended use of fertilisers and plant material, much of which is genetically modified. The vast majority of farms, both small and large, are highly mechanised and industrial. Those farmers who practise non-intensive agriculture, either by choice or on marginal lands, have little weight in food production and in the country's economy. Local know-how is often replaced by standardised industrial methods and the diversity of wild and agricultural species is decreasing.

Multinational firms are predominant actors in this scenario; they have increased their control over plant and animal production, primarily through the development of new genetic strains. Their power is moreover equivalent to that of governments.

Agrimonde 1: Feeding the Planet by Preserving Ecosystems

A Crisis Scenario

Three simultaneous crises have favoured the emergence of the Agrimonde 1 scenario:

- An acceleration of climate change, the effects of which are clearly felt from the beginning of the 2010s.
- The multiplication of food crises, from the end of the 2000s to the end of the 2010s, with highly volatile agricultural prices.
- An energy crisis, with pressure causing fossil fuel prices to rise and to be highly volatile; this situation ends only towards 2040 when the development of substitution energies brings relief.

Development through Agriculture Drives the Global Economy and Curbs Rural Exodus

From 2000 to 2050, global economic growth has been driven by the growth of the developing economies, itself strongly supported by the growth of the agricultural and agri-food sectors. The rural exodus in developing countries has slowed down owing to the economic growth of rural areas. Apart from the spread of ecological intensification practices, an infrastructure of regional planning and supply chains has been put in place in these economies: transport, storage, and industrial processing capacities, as well as services in health, education, training, and so on. The necessary investments have been made possible by improved income in rural areas. This is a result of the development of rural employment, better distribution of added value throughout supply chains, and the pooling of resources in various forms of cooperation. Public transfers implemented at national level, along with international aid for development, have been determining factors for initiating and securing investments. This massive aid was one of the answers, in the late 2010s, to the multiplication of periods of threatening food crises (linked primarily to price volatility). These situations resulted in an acceleration of inter-regional migration, especially towards the richest countries which experienced increasing difficulties in containing and managing this influx.

In certain regions the agricultural and agri-food sectors have grown significantly in the peri-urban and even urban areas around large cities, especially in SSA and ASIA where the challenge of ecological intensification has been linked to improved use of matter and energy flows from the urban system. Innovative forms of urbanisation have thus been developed.

Urbanisation nevertheless continues in all countries that have experienced strong demographic pressure. It sometimes encroaches on the best agricultural lands, despite the development of agriculture and agri-food activities in peri-urban and even urban areas, and despite efforts to densify cities in countries that were emergent in 2000 (mainly China, India and Brazil). These efforts have taken the form of strong regional planning policies to limit the artificialisation of the land and to deal with the energy crisis from the 2020s. In those instances where it has been unavoidable, the gradual conversion of the best lands has resulted essentially in a limitation of the capacity to increase yields, by pushing cultivated lands towards areas with a lower agronomic potential.

The development of the agricultural and agri-food sector has moreover strongly limited the growth of economically-driven emigration in the 2010s, contrary to expectations of an explosion of emigration due to increasingly numerous periods of food crisis. The networks of relations spawned by diasporas have nevertheless been developed and reinforced, and currently play a major part in the dynamics of innovation, knowledge dissemination and development of investment capacities, in all the economic sectors, including agriculture.

Rapid Growth of Agricultural Trade, Regulated to Ensure Sustainable Food Security

Between 2000 and 2050 international regulations have changed considerably, especially as regards food trade. In 2050 trade in foodstuffs is regulated by the United Nations Organization for Food Security (UNOFS) which, to fulfil its mission of guaranteeing food security, applies rules to avoid distortions in competition. These rules however provide for significant exceptions: (1) to enable the least productive countries to develop a local market, and (2) to take environmental issues into account. The UNOFS moreover has to ensure stocks and trade management to protect countries that are highly dependent on food imports, from threats to their supplies. The negotiations leading up to this system were initiated consequent to the food crises which multiplied from 2008 until 2020. They benefited from the effort to coordinate public policies in the environmental field (climate change and biodiversity, in particular). The active participation of environmental and development NGOs in these negotiations, alongside professional agricultural and agri-food organisations, was probably decisive in making agreement between states possible.

The tendency for agricultural prices to decrease in real terms, characteristic of the twentieth century, has ended in a huge increase in the effective demand, resulting from high population growth rates coupled with the economic take-off of countries of the South. Faced with this new situation, the regulation of markets has also aimed to avoid price volatility which was very strong at the beginning of the century and largely responsible for food crises. This action has been based primarily on the organisation of regional and global stocks, and on the strong regulation of the future market consequent to the reshaping of the international financial system. Moreover, small farmers have been able to benefit from price increases rather than suffering from them, by being increasingly integrated, either individually or collectively, in agri-food supply chains.

Agricultural Knowledge, Research & Development and Training: an Organisation Targeting Ecological Intensification

Between 2000 and 2050 research, training and development systems in the agricultural and environmental fields have produced and diffused innovations for ecological intensification. These innovations are partially specific to local agriculture but also benefit from more generic scientific breakthroughs. Many challenges that research has had to meet are related to the ability to articulate research on products—fairly typical of the green revolution—to research on systems. Innovation is organised in an interactive and often participatory way, to promote the diversity of local know-how of a variety of actors involved (farmers, other users of natural resources, NGOs, processors, etc.). This innovation effort on local, regional and global scales promotes diversity while capitalising on it and pooling it. In this respect the emergence of highly internationalised epistemic communities and communities of practice in research and ecosystem management has been decisive. The share of public research has been large in this effort, and limits have had to be placed on the appropriation of knowledge, to preserve the public nature of certain innovations and the dynamics of scientific accumulation.

Agricultural and Food Policies Strongly Linked to Environmental and Rural Development Policies

Innovations in and transformations of food and agricultural systems have been accompanied by national and regional food and agricultural policies. These have boosted the creation of regional markets in less productive agricultural areas, by mobilising a variety of tools accepted in the framework of the UNOFS, especially access to credit and insurance. In many countries these agricultural policies have also concerned land management, to enable growing rural populations to participate in agricultural development. This has implied strong intervention by states, especially for the allocation of newly cultivated land. It has also been based on innovations in terms of land ownership and tenure for agri-silvi-pastoral systems with multiple uses.

Development policies, inspired by regional development policies implemented from the late twentieth century in the European Union, contribute to structuring local and sector-specific food and agricultural systems in the form of clusters comprising processing, distribution and agri-supplies, as well as research, training and consultancy in the field of ecological intensification. In the richest countries, but also in a growing number of emergent countries, public funds are released not for production but for ecosystem management, to promote the multifunctionality of agriculture and to remunerate environmental services. These trends in agricultural policies appeared quite naturally in the late 2000s, triggered by the necessary renewal of agricultural support policies.

In many countries in all the regions, nutritional policies have been determining factors. They have been designed to promote change in consumer behaviours, and have encouraged the implementation of food product segmentation strategies in the agricultural and agri-food sectors. These policies have continued to emphasise communication towards the population, but have also proved to be highly restrictive for firms as regards information and communication on nutrition.

Energy: Technological Breakthroughs to Promote Sustainable Development

The growing scarcity of fossil fuels and the need to reduce greenhouse gas emissions have impacted on demand and renewed the energy offer by way of major investments in energy management, renewable energies and the fuel cell. The emphasis is on opportunities for distributed and decentralised energy production, waste recycling and by-products. Energy price increases at the beginning of the century triggered the search for energy autonomy on farms. It is within this framework, integrated into production as far as possible, that most of the production of agro-fuels in the world is developed.

At the beginning of the century most biofuel production was organised by oil companies, which turned towards second-generation biofuels in around 2020. The necessity for heavy capital investments, especially in R&D, explains the concentration of this sector, in which only the large agri-food or oil companies have invested. While developing countries have not really been able to join in, certain emergent countries have tried to grasp the opportunities linked to the exploitation of seaweed for energy production, especially in ASIA. No global market for first-generation biofuels emerged, as the only large producers were the US and Brazil, whose production catered essentially for the domestic market. Consequent to the energy crisis, which from 2020 seriously threatened the development of SSA, in particular, renewable energies were developed on a decentralised basis.

Environment and Global Changes: a Set of Challenges Favouring the Choice of Ecological Intensification

The diffusion of ecological intensification technologies throughout the first half of the century was motivated by a twofold necessity: coping with global changes, and ensuring sustainable productivity gains on the land. Ecological intensification technologies have made it possible to minimise the environmental impact of agricultural practices, through the reduction of greenhouse gas emissions and through other dimensions of the environment (water, biodiversity and the soil). They have likewise served to make production more resistant to setbacks, through the reintroduction of greater domestic biodiversity. The acceleration of climate change in the beginning of the century was a determining incentive for technological change in agriculture, in energy production and in sectors which are heavily energy-dependent.

The impacts of agricultural and food systems on ecosystems have been closely linked to the growth of cultivated areas, to the detriment of natural and semi-natural ecosystems. As a result, deforestation in Amazonia and the Congo Basin, which was intense in 2000, has not been entirely brought to a stop despite efforts at intensification. These efforts would have required extensive planning capacity in the management of these areas, which was sometimes lacking, especially in SSA. High prices and demand for food have created pressure to convert natural and semi-natural areas. Biodiversity conservation has therefore gone hand-in-hand with a capacity for innovation and for developing production systems that are compatible with the preservation of a rich biodiversity and ecological infrastructure, especially for farmlands reclaimed from former forests. The role of agro-forestry systems has been particularly important in this respect, even if in 2050 many environmental NGOs point to a major loss of biodiversity that justifies demands for more protected areas for wild biodiversity.

Sustainable Food and Diversity of Diets

In 2050, diets in the various regions of the world have converged as regards calorie intake; they are situated at a mean availability of about 3,000 kcal/cap/day in each region. Cultural particularities have nevertheless maintained some diversity in the distribution of the various food sources. Increasing incomes have not led to a convergence of diets towards Western diets. Even though in certain regions, especially SSA, food consumption trends have been a result of economic development, they have also stemmed from behavioural changes in most regions. For instance, in a region like OECD, the mean calorie consumption has declined from 4,000 to 3,000 kcal/cap/day. This steep downward trend is the result of less wastage by users and catering systems, and more effective nutrition policies. The evolution of diets towards lower calorie intake has occurred without any drop in household expenditure on food. This can be explained by an increase in the added value of the products consumed (diversity, quality, etc.). The decrease of mean calorie availability also reflects less inequality in calorie intake within regions or countries.

The increase in the proportion of animal calories in diet was a strong trend in 2000 but has developed differently in each region, especially as regards the proportion of red meat and white meat. In SSA there has been an increase in the consumption of white meat, mainly, and of proteins from aquatic sources. In ASIA diets have remained strongly linked to cultural preferences. In China the share of red meat in diets has grown in proportion to increasing incomes, although fresh- and sea-water fish is still a regular part of diets. In India, on the other hand, a large proportion of the population has remained vegetarian. The share of aquatic products in most diets throughout the world has increased substantially, although this has not replaced other sources of protein. The growth in the consumption of these products has been strongest in ASIA and LAM, where aquatic production has developed considerably, and to a lesser extent in SSA.

Processed products from aquaculture still have a large market, although the growth of their share of the total has declined. The maintenance of diversity in diets also helps to solve problems of micronutrient deficiencies, primarily through the consumption of fruit and vegetables and the variety of products. The fast growth of the proportion of processed products compared to raw products, recorded at the beginning of the century, has levelled off. This is a symptom of the diversification of food systems. It also stems from regulations which have placed strong constraints on agri-food companies' information and communication on nutrition, in rich countries, encouraging them to limit the degree of product processing while continuing to sell innovative products in terms of practicality and variety. This is moreover another factor explaining the reduction of waste by consumers.

Diversified Agricultural and Food Systems

From 2000 to 2050, the agri-industrial model, initially clearly dominant, has merged with more local food and agricultural systems based on short circuits and on the diversity of small and medium-sized farms and processing enterprises, especially in developing countries. In these countries the establishment of agri-food multinationals and mass retailers has been attended by various partnership and innovation strategies between firms and local agriculture. The tendency towards standardisation, internationalisation and concentration around a limited number of multinational firms has therefore declined. This change has been facilitated by national and regional strategies to ensure food security, and by the considerable impact of Corporate Social Responsibility on large firms' strategies. The agri-food sector has been strongly affected by this movement because consumers in rich countries have proved to be more and more concerned about food issues, due to the spread of the sustainable food concept (the threat of an obesity epidemic has resulted in the proliferation of information and education campaigns on nutrition, which have also affected the upsurge of "environmental awareness" by associating human health with environmental health). Moreover, hunger riots have made a strong impression on consumers, who pressurise agri-food firms, often via NGOs and consumer organisations, to assume their responsibilities in economic development and the reduction of malnutrition, as well as in the struggle against obesity.

Owing to diversification strategies, the trend towards specialisation on a limited number of basic products has been reversed. The diversification of supply chains, actors and products has been complementary to the ecological intensification of agriculture, by enabling the diversity of products adapted to local ecological conditions to find markets. In all regions, small and medium-sized agricultural and food firms have developed a real capacity for innovation and job creation (often in rural areas), capable of competing and cooperating with the dominant actors. The expression of this innovation capacity has however been possible in 2050 only by maintaining a diversity of modes of governance of supply chains and economic sectors (cooperatives, partners, and not only share-holders), as well as territories.

In 2050 production systems have become widely diversified and adapted to local ecological conditions. They incorporate a set of agri-silvi-pastoral and environmental functions, and are integrated into territorial systems which include the development of transport, education and health infrastructures, energy production, processing and commercialisation. Where relevant, they may be articulated to the production of biomaterials from silviculture, or to the urban systems to which they are related. Soilless crop systems have not played an important role, primarily because most farms in developing countries have been unable to afford the required investments.

Prices of aquatic products remain high because the supply has been unable to meet the demand even though it has continued to grow through aquaculture. Fish supplies have declined as a result of international efforts to manage marine resources, and aquaculture has encountered problems of space and conflicts over use, which have limited its development in fresh water. Expectations have therefore been concentrated on marine aquaculture. This development has nevertheless had to cope with a set of problems that research and innovation have helped to solve:

- the domestication of the most valued species,
- conflicts over the use of space, even in the sea,
- the difficulty of procuring animal food of marine origin, non-substitutable by food of land origin, which is too expensive,
- potential environmental impacts.

This marine aquaculture has developed mainly in specific tropical areas where conditions and the extent of demand are favourable. These evolutions have mainly taken the form of intensification systems on the shoreline. An effort has been made to integrate them effectively into food production and processing systems, with a specific focus on traceability and low environmental impacts. Research on the environmental impacts of large marine aquaculture farms has also been undertaken. Knowledge on and production of seaweed has furthermore become an important area of research and innovation, and the production of biomass for energy or food, in seaweed aquaculture, has found large markets.

Chapter 11
Scenario-Based Insight into Food Behaviours, Technological Options and Trade

Jean-Marc Chaumet, Francis Delpeuch, Gérard Ghersi, Jean-Louis Rastoin, Bernard Hubert, Tévécia Ronzon and Michel Petit

The scenarios presented in the preceding chapter represent two of the many possible future trajectories of the world. Their merit is primarily to highlight the extent of changes that can usefully be considered for the world's food and agricultural systems over the next four decades, up to 2050. These changes are not unrelated to one another; future evolutions of consumption and production will interact, irrespective of the scenario. It is therefore necessary to foresee all of these changes coherently and comprehensively, taking all their technical, social, cultural, economic and political dimensions into account.

J.-M. Chaumet (✉)
Institut de l'Elevage, 147 Rue de Bercy, 75595, Paris Cedex 12, France
e-mail: Jean-Marc.Chaumet@inst-elevage.asso.fr

F. Delpeuch
IRD, BP 8967, 15 rue Abou Derr, 10000, Rabat Agdal, Maroc
e-mail: francis.delpeuch@ird.fr

G. Ghersi
Montpellier Cedex 5, France
e-mail: gerard.ghersi@univ-montp3.fr

J.-L. Rastoin
SupAgro Montpellier, UMR 1110 Moisa, 2 place Pierre-Viala,
34060, Montpellier Cedex 1, France
e-mail: rastoin@supagro.inra.fr

B. Hubert
Agropolis international, 34394, Montpellier Cedex 5, France
e-mail: bernard.hubert@avignon.inra.fr

T. Ronzon
IPTS, C/Inca Garcilaso, s/n 41092, Seville, Spain
e-mail: tevecia.ronzon@ec.europa.eu

M. Petit
IAMM, 3191 route de Mende, 34093, Montpellier Cedex 5, France
e-mail: petit@iamm.fr

S. Paillard et al. (eds.), *Agrimonde – Scenarios and Challenges for Feeding the World in 2050*,
DOI 10.1007/978-94-017-8745-1_11, © Éditions Quæ, 2014

After building two scenarios illustrating the variety of possible evolutions and their interdependencies, this chapter considers the following questions: What evolutions will be possible? Necessary? Desirable? Generally, the exploration presented in the preceding chapters does not enable us to answer these questions unequivocally. The scenarios analysed do nevertheless provide us with a valuable framework to discuss the most desirable, necessary, or feasible options.

In this respect this chapter constitutes an opening. Starting with a few important points, it uses Agrimonde scenarios to structure strategic discussion on the future of the world's food and agricultural systems. To draw the main conclusions from the foresight analysis, possible readings of scenarios are proposed, which do not necessarily reflect consensus within the Agrimonde panel. The three points of entry are: diet, technological options, and trade.

Food Behaviours in Question: Are Ruptures Plausible?

The assumptions on diet used in the Agrimonde 1 scenario are strongly based on changing trends. They suppose that sub-Saharan Africa (SSA) will experience rapid economic development providing its inhabitants with access to food that is equivalent in 2050, on average and in terms of both plant and animal calories, to that of other regions of the world. This contrasts with the situation in 2000 when it represented respectively 88 % (plant calories) and 28 % (animal calories) of the average global availability. Moreover, Agrimonde 1 assumes that there will be considerable changes in food consumption behaviours. The present section considers these major changes in behaviour.

To simplify, Agrimonde 1 foresees two types of change regarding food. First, this scenario assumes that the mean level of availability of food for consumption (in kcal) will decline in rich countries[1] without this being due to a decrease in the mean per capita income[2]. Second, it assumes that, in developing countries, an increase in mean per capita income results in an increase in food availability that is large enough to counteract risks of undernourishment but moderate enough to ensure that the nutrition transition does not result in a new progression of obesity.

The long-term projections on food consumption in the world or the main regions are generally based on a function linking food consumption to income. This is notably the case of the scenarios in the Millennium Ecosystem Assessment (MA). Global Orchestration foresees a steep increase in food consumption, especially of

[1] In any case, in the rich countries that have exceeded the annual mean currently situated at 3,000 kcal/cap/day, as in 2050 in Agrimonde 1.

[2] Mean availability for consumption is considered here to be equal to the availability and not to the ingestion of food, as available food is partially ingested and partially wasted. In terms of ingestion, humans' net energy needs are between 2,000 and 3,000 kcal/day, depending on sex, height, weight and intensity of physical activity.

meat, as a consequence of a very high global economic growth rate, especially in the economies of the regions currently qualified as emergent (MA 2005b). The possibility of a trend rupture is one of the main questions raised in the systematic analysis of assumptions on the evolution of diet presented in Chap. 5. It is therefore important here to consider the main elements of plausibility or questions concerning this rupture.

In the first section of this chapter the relationship between diet and income is described. The factors related to demographic trends that will influence this relationship in the period leading up to 2050 are discussed. In the following sections other factors liable to impact on diets and bring them closer to the Agrimonde 1 set of assumptions are explored. These factors concern the fact that the environmental and nutritional implications of food consumption are taken into account, both individually and collectively.

The Relationship between Diet and Income, and its Determinants

Income and supply permitting, the evolution of food consumption is characterised by three phases (Combris 1990). The first phase is one of quantitative growth of consumption of all foods until calorie saturation is reached. The second phase is characterised by a qualitative evolution centred on the structure of rations: dietary transition (Popkins 1993). It is based on three main factors:

- the rise of living standards in developing countries, which tends to increase food consumption and above all to bring meat consumption levels closer to those of developed countries,
- urbanisation: the proportion of urban populations in the world are steadily climbing; all population projections foresee a much higher percentage of urban dwellers in 2050[3]. This trend influences diet and modes of consumption. City-dwellers' diet is generally more varied than that of rural populations, and they consume more processed foods. More fish, fresh vegetables, meat, poultry, milk and other dairy products are consumed. However this consumption is often nutritionally unbalanced; it is too rich in calories, with too much sugar and saturated fats. Combined with less physical activity compared to rural populations, this mode of consumption often entails risks of obesity and cardio-vascular diseases. Urbanisation is also often attended by a reduction in the amount of time spent on preparing meals, an increase in the number of meals taken out of the home, and a loss of culinary skills (Latham 1997),
- the industrialisation of food systems, with consumption characterised by: increasing proportions of processed and ultra-processed foods (ready to cook or to consume) and of packaging; increase in animal products from battery breeding;

[3] Up to nearly 70 % according to the UN scenario, with a growth of urban populations from 2.8 billion in 2000 to 6.4 billion in 2050 (UN 2007).

predominance of supermarkets in retail trade and fast-food chains in the catering industry, etc. Food processing nevertheless allows for effective conservation of products, which limits waste after slaughtering or harvesting.

Finally, the third phase is characterised by no change in the macro-nutritional structure of food rations. Countries' wealth is reflected in a levelling off, from a certain threshold, of the amount of food available per person, associated not with a decrease in spending on food but with a purchase of more sophisticated food (more elaborate preparation and packaging, or food of better quality) which is therefore more expensive for the same calorie content.

This non-linearity of the function linking consumption with income is accentuated by other factors[4]. First, as elderly people need fewer calories than adults or children do, the ageing of the population entails a decrease in food consumption[5]. This effect will be considerable in 2050 due to the increase in the proportion of elderly people (over the age of 60) in the world's population. UN estimates show an increase in elderly people from 610 million in 2000 (i.e. 10 % of the population) to over two billion world-wide in 2050 (i.e. 21 % of the population) (UN 2006). All regions are concerned, including Africa, even though this continent will still have the lowest proportion of elderly people. Secondly, certain diets remain stable despite the country's development and the increase in household incomes. Japan is an example: this country has had a diet characterised since the 1960s by daily per capita food availability of about 2,800 kcal.

Agrimonde 1 as a Scenario of Reduced Waste and Sustainable Consumption

The Agrimonde 1 assumptions on food consumption could, in the spirit of this scenario, reflect an effective struggle against waste by the final consumer. Certain behavioural trends, still marginal today, may be generalised by 2050. In recent years there has been a slow but growing individual and collective awareness of the sometimes harmful effects of each individual's actions in daily life, with regard to health and the environment. A more responsible attitude, entailing sustainable behaviours, is encouraged by actors in civil society and by public authorities endeavouring to raise public awareness. A generalisation of this awareness in the general public and among policy-makers in rich and emergent countries is essential if the changes foreseen in Agrimonde 1 are effectively to occur.

[4] To be complete, it is necessary to point out a methodological limit to Agrimonde: as the quantitative model does not take into account the prices of products, it was not possible for us to examine the influence of food prices on consumers' diets. Yet changes in consumption trends in OECD countries would be compatible with a scenario in which prices, especially of meat, remained high.

[5] However, nutritional recommendations for elderly people lead to a minor reduction in the energy intake. The risk is essentially under-nutrition in case of over-reduction. In particular, as the metabolisation of proteins becomes less effective with age, the protein intake must remain at least equal to that of adult men or women, and must be constituted primarily of animal proteins which are more easily metabolised.

Many governments are implementing measures to encourage more sustainable consumption (OECD 2008). Denmark, for example, increased the price of domestic water by 150 % through various taxes, which had the effect of reducing consumption. In many countries labels showing the level of energy efficiency are now compulsory on a large number of household appliances. Italy, which chairs the "Marakech Task Force on Education for Sustainable Consumption", has set up a network of schools of the future which teach and implement social and environmental sustainability. Citizens participate directly in this consciousness-raising through non-profit organisations that promote more sustainable behaviours.

Willingness to promote more sustainable consumption is also developing through local agri-food systems which make it possible to reduce transport costs and out-of-season food production, thus reducing energy consumption and environmental impacts. One of the ways in which these principles have been implemented is through the "locavore" movement initiated in the US in 2005, which encourages consumers to buy fresh local produce in season, produced within a radius of 150–200 km.

Even though researchers in some countries are starting to focus on this issue by launching programmes on sustainable consumer behaviours, our knowledge of interactions between consumption, especially of food, and sustainable development is still fragmented. For example, comparison of the environmental impacts of very localised food systems on the one hand and highly internationalised systems on the other, raises complex questions that are still largely unexplored—even though they seem to be essential to assess the various options for the future.

Agrimonde GO and the Fast Growth of Nutrition-Related Non-Communicable Diseases

Nutrition-related non-communicable diseases are gradually becoming a topical subject and are a cause for concern among citizens and governments alike, in all regions of the world. The rapid increase in health expenditure and the economic and social costs of these diseases are worrying (Box 11.1). Agrimonde GO, as a trend-based scenario on diet, shows the risk of an explosion of these costs.

In 2005, 1.3 billion adults in the world were overweight, of which 400 million were obese. If recent trends continue, in 2030 these figures could reach 3 billion and 1 billion, respectively (Kelly et al. 2008). In a recent study it was found that 17 % of French people are obese and 32 % overweight (USEN 2007). In the US, the proportion of overweight adults rose from 47 % in 1980 to 64 % in 2000, and obese adults from 15 to 31 %[6]. During the same period, the proportion of overweight children and adolescents tripled. At the current pace, 20 % of the French population could be obese by 2020 (INSERM, TNS Healthcare, Roche 2003), and 40 % of the US population (Ruhm 2007).

[6] NHANES, Data on the Prevalence of Overweight and Obesity among Adults.

The problem of overweight and obesity, sometimes qualified as an epidemic, is also spreading increasingly in developing countries, where 800 million people are overweight and close to 200 million obese (Kelly et al. 2008). This trend is due to the spread of the Western model of consumption and to public policies, as food aid has often consisted of obesogenic food (Delpeuch et al. 2009). As a result, 18% of the Uruguayan and 25% of the Egyptian population are reportedly obese[7]. As obesity co-exists with chronic under-nutrition and micronutrient deficiencies, most developing countries are experiencing a "double burden" (WHO and FAO 2003). Moreover, even though the causes are not entirely elucidated, the impact of diabetes and hypertension on overweight people is felt more rapidly in developing countries than in rich countries, as people in the former have a greater capacity to store fats (Basdevant 2003).

Box 11.1—Overweight and obesity: definitions, causes, risk factors and associated disorders

Overweight and obesity are defined as abnormal or excessive fat accumulation that may impair health. Body mass index (BMI) is a simple index of weight-for-height that is commonly used in classifying overweight and obesity in adult populations and individuals. It is defined as the weight in kilograms divided by the square of the height in meters (kg/m^2). BMI provides the most useful population-level measure of overweight and obesity as it is the same for both sexes and for all ages of adults. However, it should be considered as a rough guide because it may not correspond to the same degree of fatness in different individuals.

The World Health Organization (WHO) defines "overweight" as a BMI equal to or more than 25, and "obesity" as a BMI equal to or more than 30.

The fundamental cause of obesity and overweight is an energy imbalance between calories consumed on one hand, and calories expended on the other hand. Global increases in overweight and obesity are attributable to a number of factors including:

- a global shift in diet towards increased intake of energy-dense foods that are high in fat and sugars but low in vitamins, minerals and other micronutrients; and
- a trend towards decreased physical activity due to the increasingly sedentary nature of many forms of work, changing modes of transportation, and increasing urbanisation.

But other factors may also be involved:

- genetic factors: even if many uncertainties remain as to the part that genes play in the occurrence of obesity, some researchers think that they do at least partially explain the variance of BMI. The modalities of interaction of genetic factors with various environmental factors (especially nutrition) have not yet been elucidated but several genes seem to be involved.

[7] World Health Statistics, WHO.

- psychological factors: certain psychological states such as depression or stress can be conducive to weight gain. It is difficult to measure the consequences of these conditions, for obesity in itself may contribute to the development of certain psychological features.
- social and cultural factors: in all rich countries, a relationship between poverty and obesity can be observed. The explanation may be bad diet and inequality of access to healthcare.

Overweight and obesity have serious effects on a person's health and risks increase in proportion to BMI. A high BMI is a high risk factor for chronic ailments such as:

- cardio-vascular diseases (mainly heart-attacks and strokes) which are already the main cause of death in the world (17 million deaths per annum),
- Type-2 diabetes which has rapidly grown to the proportions of a global epidemic. The WHO predicts that deaths due to diabetes will increase by over 50 % in the world over the next ten years,
- muscular-skeletal diseases, especially arthritis,
- certain cancers (endometrium, breast and colon).

Source: WHO website <www.who.int>

Agrimonde 1 Illustrating an Effective Struggle against Nutrition-Related Diseases

The Agrimonde 1 assumptions and the behavioural changes that they imply form a scenario in which the struggle against these diseases is not only a top priority but is also successful, along with the fight against hunger and undernourishment. The extent of undernourishment is unfortunately still disastrous and actually increasing in the early twenty-first century, both globally and at country level, including in the OECD, despite various policies and measures to eradicate it.

In recent years numerous political measures have been taken to prevent nutrition-related diseases, which could lead to the healthier diets foreseen in the Agrimonde 1 scenario. Examples include France, the US, the UK, Canada, Tunisia and China. France has set up a National Nutrition and Health Programme (PNNS) covering the periods 2001–2006 and then 2006–2010. The second version of this programme has three objectives: first, prevention through education in nutrition; second, early screening and treatment for nutrition-related disorders (obesity, under-nourishment); and, third, targeting underprivileged populations. These objectives are associated with quantitative indicators for a five-year term (−20 % prevalence of overweight, −25 % "low consumers of fruit and vegetables[8]", −5 % moderate cholesterolaemia, +25 % individuals practising physical

[8] That is, consumers who eat less fruit and vegetables than the recommended daily intake.

exercise) organised according to 9 references and regularly monitored (Ministère de la santé et des solidarities 2006).

The US launched the CARE (Communication, Action, Research and Evaluation) project consisting of various types of action: providing information and educating people on health problems; implementing programmes (especially in schools) and carrying out research to further understanding of the causes and treatments of obesity (Combrie et al. 2005).

At the beginning of 2008, the UK government launched a campaign against obesity, focused on nutritional education of children and adolescents.

Canada published Canada's Food Guide and in 2007 set up tax incentives for families whose children practised certain sports.

As early as 1995 Tunisia launched an original programme in the context of the time, the National Food and Nutrition Programme, to "sustainably achieve Tunisians' nutritional well-being" by integrating a nutrition section into development policies.

China has started to take measures by publishing a National Plan of Action for Nutrition, and Dietary Guidelines for Chinese Residents (D'Arcy et al. 2006). A national system of systematic prevention of certain diseases (high blood pressure, Type-2 diabetes, and certain cancers) has also been set up recently in China.

The nutritional situation of populations is also a subject of concern in inter-governmental organisations. In 2003 the WHO and the Food and Agriculture Organization of the United Nations (FAO) published a report entitled "Diet, Nutrition and the Prevention of Chronic Diseases" (WHO and FAO 2003) and in 2004, on the basis of this report, the World Health Assembly adopted a "global strategy on food, physical exercise and health".

Yet the policies implemented by governments to prevent and combat nutrition-related non-communicable diseases seem to have been relatively ineffective until now. For instance, in 2001 the French government launched the PNNS with the aim of reducing the prevalence of obesity in France by 2005. Although obesity and overweight in children in the 7–9 age-group seem to have remained stable (Salanave et al. 2008), a study shows that in 2009 adult obesity had progressed in France, compared to 2000 (INSERM, TNS Healthcare, Roche 2009). The impact of public nutrition programmes naturally depends on government grants. The second PNNS has a budget of only 47 million €, which is very little compared to the 5 billion spent annually on advertising by large agri-food firms. In Tunisia, the National Institute for Nutrition and Food Technology, charged with implementing the National Food and Nutrition Programme, suffers from a lack of funds to the extent that its intervention has been limited to surveys on nutrition, classes in schools, and radio programmes to raise public awareness (Dekhili 2004). Here again, there is a discrepancy between laudable intentions and achievements which remain limited and therefore of little effect on public health.

Policy evaluation is difficult as it usually concerns a quantitative review of the actions carried out or the description of processes. Evaluations rarely study changes of behaviour or make use of medical indicators (Combris et al. 2008). Over and above this difficulty pertaining to evaluation, the policy implementation encounters many difficulties. For example, it is acknowledged that the struggle against obesity is confronted with a large number and variety of factors that can explain this disease (Box 11.1), and that it therefore requires policies at several levels of intervention (Dériot 2005). Moreover, the agri-food industries also weigh on the behaviour of consumers and their messages are not always compatible with the objectives of public health.

More generally, public interventions in developed countries, mainly scarce until now, have focused primarily on information, education and communication. They do not seem to have had any significant effect on the consumption habits of citizens, the majority of whom have not altered their diet. Actions based exclusively on public awareness and education do not seem sufficient to trigger changes in behaviour; a transformation of the currently obesogenic environment must be one of policy-makers' targets (Swinburn and Egger 2002). In developed countries we are witnessing a slowly growing awareness of the consequences of diet on health, and therefore a relatively slow change in this respect. This awareness will be very difficult to achieve in developing countries where social success is often associated with abundance and a sedentary lifestyle, and where very little nutritional information is disseminated.

One of the challenges of the Agrimonde 1 scenario is therefore to find more effective forms of action to trigger the change foreseen in food consumption trends. To that end, decisive progress in our knowledge is required, especially to further understanding of the complexity of food behaviours and their relationship with health and the environment, and to identify what is likely to change them.

Options for Ecological Intensification: Technical, Social and Territorial Changes

Today the concept of ecological intensification essentially refers to technical options to develop, rather than a prescribed set of processes that can be directly transferred. It may therefore very quickly appear that, as in fields other than agriculture, these so-called technical options also encompass social, economic, spatial and political options which are not incidental and have probably not been sufficiently explored. We do however know enough about the choices that have accompanied the rationalisation process ("modernisation") of North American and European agriculture, since their effects are now visible, even if they have not all been clearly expressed nor deliberately intentional. This knowledge enables us to clarify the conditions required for implementing a particular option. With this background knowledge and based on Michel Griffon's book *Nourrir la planète*, the Agrimonde panel chose to explore the possibilities of so-called ecological intensification (Conway 1998) and the different realities that it can encompass, depending on the region or ecosystem considered (Griffon 2006).

Ecological Intensification as a Technical Option

First, let us revert to ecological intensification, as such. It can be defined as a technical alternative to the steady development of an agriculture based on the substitution of labour by capital, by means of mechanisation and heavy consumption of fossil fuels, as well as an artificialisation of production conditions aimed at freeing farmers from the constraints of natural processes through the use of manufactured inputs (fertilisers, pesticides, animal fodder, etc.), genetic selection of plant varieties (distinct, homogeneous and stable) and improved animal breeds, etc. The idea of ecological intensification is to revise some of these choices, and to conceive of a type of agronomy that is closer to steering ecological processes than to an attempt to control the production process as tightly as possible. This new agenda also implies: lower consumption of fossil fuels; better use of the soil's ability to mobilise organic matter (by associating or sequencing certain crops and using new tillage techniques); integrated pest management (biological control organisms, mixing species and varieties, field organisation, crop rotation, etc.); greater flexibility and adaptability of the species cultivated or bred; better disease resistance by relying on diversified populations and not on standardised genomes, etc. Of course, none of this has been well established yet and it all requires research and experimentation by researchers as well as farmers. For the advocates of ecological intensification this does not mean reverting to an archaic type of agriculture; on the contrary, the aim is to use modern techniques to attain its objectives: marker-assisted selection, biotechnologies, tillage techniques, matching of technological recommendations with the micro-local ecological conditions, mechanisation as well as animal traction, etc. It is difficult today—without an accurate inventory of available test results—to know what yields can be obtained. The Agrimonde 1 scenario proposes targeted yield objectives, beneath which this type of approach remains utopian. The options are open, the principles are starting to become known, the environmental imperatives are becoming clearer, the threat of climate change is emerging, knowledge is already available to some—and not only in the research community. Certain national programmes (such as the SYSTERRA programme launched by the National Research Agency in France) or international programmes (like some of the Challenge Programmes of the CGIAR [Consultative Group on International Agricultural Research] or the International Assessment of Agricultural Knowledge, Science and Technology for Development (IAASTD 2009)) provide strong incentives to strengthen research in these directions, especially since such new technical options have considerable social, economic and spatial dimensions.

Ecological Intensification as an Option for Social and Spatial Organisation

The question of the agrarian frontier, which seemed to have been forgotten somewhat, is back on the agenda. To the first frontier, that of forest clearing and the cultivation of "virgin lands", well-known since the Neolithic, a second frontier has inexorably been added over the past century or more, that of urban and infrastructure development. Here, regulations and the land market set the rules, and agricultural profits are rarely able to withstand other types of speculation or decisions made in the general interest. Finally, new environmental and social issues suggest the existence of a third frontier, within the agricultural world and based on the way of conceiving of crop and stock farming techniques.

The question of urban development was examined closely for the regions of Asia (ASIA) and Middle East—North Africa (MENA) in the Agrimonde 1 scenario. In 2050 these two regions combine the characteristics of having attained the limits of their cultivation potential and having experienced a population explosion, probably a driver of intense urbanisation both in rural areas and around cities. The result will be exacerbated competition between urban areas and agricultural areas for land and natural resources (water, firewood, etc.)—not to mention the potential social tensions. It is therefore probably time to reason differently as regards what urban and peri-urban agriculture might be, other than a rival—vanquished from the start—of residential or industrial expansion. It could contribute to dotting urban landscapes with spaces devoted to production, although this would trigger issues on the rehabilitation and conservation of land polluted by refuse and installations, or simply subjected to the effects and consequences of various urban activities. Questions would likewise be raised on the organisational dimensions of the commodification and retailing of food, by the creation of short circuits that exclude most of the usual operators involved in their commercialisation. Such short circuits could result from the growth of "traditional" neighbourhood markets linking urban consumers to local producers. Another possible development already exists in Europe, certain North American cities and in emergent countries such as Brazil, consisting of new forms of direct agreement between consumer groups and producers for the supply of fresh produce to households. In these agreements the terms of reference are set for the entire production cycle and prices are guaranteed *ex ante*.

The first frontier, that of clearing land and control over nature, is conceived of in two ways which differ essentially as regards the capacity to preserve ecosystems and their natural resources, and the way in which this issue is addressed in terms of spatial organisation.

On the one hand, those who defend the inviolability of this frontier do so today in the name of the local dimension of global environmental issues. Biodiversity, protection of the soil, preservation of water resources and their quality, and the slowing down of climate change by carbon storage, are all reasons to stop clearing the world's large forests (Amazon, Congo Basin, etc.) and draining wetlands to use them for agriculture[9]. Those who defend these ideas believe that it is necessary to intensify production without changing the frontier. The aim is to protect sanctuaries and reserves and, if possible, to promote a "cleaner" agriculture, such as the eco-agriculture advocated by the IUCN (International Union for Conservation of Nature) for example (Mc Neely and Scherr 2003). From this point of view, ecosystem services will partly offset technical excesses (biological and abiotic contaminants),

[9] Unless there is compensation, as in the US in the case of mitigation banking (www.mitigationbanking.org).

should these occur in agricultural areas, and provide guarantees on some major global balances in view of the technological challenge of feeding the world's 9 billion inhabitants in 2050. Even if the idea today is not to present this frontier as a land reserve, its use will become an issue in the case of the taiga when, with the disappearance of the permafrost, there will be strong pressure to cultivate or convert to pasture these lands abounding in organic matter stored since the last glaciations, and emitting CO_2.

For proponents of Californian School Agro-ecology[10], the answer requires the identification of a third frontier, within the agricultural world. This implies the need to revise the very conceptions and practices of agricultural activity (Altieri 1987). They argue for a different conception of the world to the current conception that divides it into towns and cities where most of the population is concentrated, rural areas devoted to agricultural production or forests, and natural areas guaranteeing a supply of nature and the preservation of the earth's main balances. It is proposed that we conceive of a diversity and complementarity of forms of agriculture, based on knowledge and practices, technical models and channels of commercialisation singularised according to the production, geographic situation, form of public action, cultural values, and engagement in an individual or collective mode. Such forms of agriculture could be present in urban areas and forests and thus introduce original ecological mosaics, inventing modes of production inspired more by ecology than by chemicals. They could rehabilitate the diversity and variability of plant and animal genetic material, as well as of the soil, and make good use of slopes, hollows, selvedge, and plant and forest stratas for production purposes, while respecting these areas and guaranteeing their resilience.

In the Agrimonde 1 scenario we did not wish to opt for one of these two conceptions *a priori*. We preferred to compare each option with the regional context under consideration, and then to choose the one that seemed to correspond best to the fundamental principles of the scenario. Thus, depending on the region, the application of the concept of ecological intensification takes different forms, based on one of the two options presented above. In certain cases the Agrimonde GO scenario takes the opposite stand to that of the Agrimonde 1 scenario. It was however the analysis of these contrasting but justified situations that made it possible to explore the notion of ecological intensification more thoroughly, along with its environmental, economic and social implications (Chap. 9).

The "Segregationist" Model: Intensifying Production Systems and Protecting Natural Areas

The first model, that could be qualified as "segregationist", thus separates that which can be cultivated from that which must not be cultivated from the point

[10] Amongst others, the Universities of Berkeley (Department of Environmental Science, Policy and Management), Santa Cruz (Center for Agroecology and Sustainable Food System), Santa Clara (Environmental Studies Institute), and Davis (Agricultural Sustainability Institute).

of view of environmental protection, but in which "natural" processes must nevertheless be managed. This situation is clearly illustrated in the Agrimonde GO scenario by choices made for the Latin America region (LAM), which can be summed up as a stabilisation of forest areas (loss of only 1 % of the forest's surface area from 2000) offset by high yield gains in cultivated areas (doubled in 50 years). In the OECD-1990 region (OECD), the Agrimonde 1 scenario presents a more modulated version of this segregationist model. The reforestation shown between 2000 and 2050 shows the will to preserve natural areas; however, due to reduced possibilities for intensifying cropping systems, pastures are gradually turned into cultivated areas.

The so-called segregationist model's management of natural areas has already inspired an abundant literature—within the scientific community as well as NGOs involved in advocating these proposals—on approaches based on ecosystem management concepts (the details of the different variants will not be presented here). These approaches aim to steer the dynamics of a world to which humans are foreign, but that they must nevertheless preserve to protect the planet on which they are living. In artificially cultivated or wooded areas devoted to human activities, damage to natural processes must be limited by penalising polluting practices and encouraging those favourable to biodiversity, soil conservation, protection of the quality of water resources, and so on. This variant of ecological intensification in traditional farming areas requires innovations in less environmentally harmful agricultural practices. Proposals usually concern new pest-control or soil conservation techniques in which biotechnologies, precision agriculture and so on can play a large part—especially if the performance criteria are those currently used, with mainly techno-economic indicators albeit under the constraints of environmental criteria. In any case, environmental challenges are considered as being met elsewhere, in the spaces devoted to them, that is, reserves, corridors and natural areas which fulfil this function for the entire planet and are justified by the services that the ecosystems concerned render to humankind.

For as long as performance criteria remain unchanged (i.e. yields, weight gains, labour productivity, etc.), albeit under certain environmental constraints, they will favour those farms that apply the recommended technologies the most efficiently, and that are able to make big investments and to benefit from appropriate technical support. In the example of LAM in the Agrimonde GO scenario, this variant of ecological intensification remains fully compatible with the scenario's rationale: pursuit of maximum economic growth, evaluated with classical economic performance criteria. The compatibility of this variant in which priority is given to agricultural development and equity in the Agrimonde 1 scenario is however questioned. We know that the farms which readily adopt these technologies are those that have considerable means (or aid) for investment and appropriate technical support, i.e. those that derive the most benefits from comparative advantages of technical innovations in a competitive system. In this type of model, so-called "commercial" farms that are sufficiently large to be efficient, and those in the process of becoming so (through investment and technology acquisition), are distinguished from others which will never become like that and are destined to disappear or to be treated "socially".

Thus, in the Agrimonde 1 scenario, the form of ecological intensification closest to the segregationist model is found in the OECD region, where a majority of farms have access to financial and technical support to adopt the technologies required. The former Soviet Union could also move towards a similar type of agriculture on its cultivated lands, due to a lack of manpower. However, in this region the frontier between natural areas and productive areas is less tangible. Forests become the main biofuel production area, and the frontier marked by the permafrost moves north to make way for forms of agro-pastoralism. In the other regions the main hindrance to the generalisation of this type of model lies in the possibilities of access to technology, as the lack of financial, human and sometimes land resources automatically excludes poorer farmers from this model. Moreover, the capacities to administer natural reserves and other ecological corridors, as well as the governance required by a model in which the pioneering front of natural areas has been completely stopped, may constitute another major impediment to the adoption of this model in these regions.

The "Integrationist" Model: Combining the Ecological and Productive Functions of Agro-Ecosystems within the Same Territory

The mode of ecological intensification opted for in the Agrimonde 1 scenario in the MENA, SSA, LAM and ASIA regions is thus closer to the second model, which could be qualified as an "integration" model. The approach is different though, for this model is based on the combination, in the same territory, of different types of productive systems adapted to the ecosystems constituting the territory, in such a way as to maintain it in the form of a mosaic of ecosystems providing various services (purifying and regulating water resources, soil conservation, maintenance of landscape structures and biodiversity, carbon storage, etc.). This involves different types of farming (livestock, forestry, crops, etc.) in the same territory, on the same farm or on different farms, overlapping to differing degrees.

The agricultural performance criteria are no longer limited to techno-economic indicators such as productivity of work per hectare on the scale of a farm. They encompass a range of indicators, on the scale of a territory, which provide information on the efficiency of agricultural practices as regards water quality, biodiversity and soil quality conservation, as much as on commodity products. These criteria may be defined more in terms of the way in which a hectare of cultivated land can provide these various services in a sustainable way. In this schema the types of productive systems described above are no longer exclusive; they are complementary in so far as they allow for efficient management of the diversity of the ecosystems involved. The Agrimonde 1 scenario is a fine illustration of this. For instance, very generally speaking, in LAM, forests are no longer devoted to protection or to clearing for land use, but to intermediate forms corresponding to various agro-forestry models. In ASIA humid areas are not all drained; rather, they are valued as a source of grazing land in dry seasons or for combined agricultural and aquaculture projects. In MENA and in SSA, rangelands with low forage productivity become key elements in grazing routes that use a diversity of environments and biological corridors enabling the fauna and flora to circulate. The same applies to hedges, small woods and orchards, habitats for many crop auxiliaries and coarse substances that preserve the soil and low-lying vegetation from the effects of wind and rain. In the Agrimonde 1 scenario, farms with a low level of efficiency in terms of exclusively techno-economic criteria used at the beginning of the period (2000–2010) play an important role in this respect in 2050. They make the multifunctionality of agriculture fully meaningful, that is, not only a farming activity that provides goods and services apart from agricultural goods, whether for food or otherwise, but also one of the activities practised in a territory by some of the households living there. In this sense it is both the territory and the households that are multifunctional, as agriculture as such represents only one of these functions.

Further Exploration of the Concept of Ecological Intensification

In this sense, the Agrimonde 1 scenario integrates a change of viewpoint on the multifunctionality of agriculture (Caron et al. 2008), which was regarded as essential by the recommendations of the IAASTD (IAASTD 2009) and by the World Bank 2008 report on agricultural issues (World Bank 2008). One of the first tasks to make this multifunctionality meaningful would consist in producing performance criteria to evaluate the accomplishment of the different functions, if only to frame them politically and to administer them, in addition to remunerating them. We would then see that in such a schema the different types of agriculture mentioned above complete one another rather than having to conform to a single model.

Finally, in both cases, but more so in the integration model, the question arises of the real capacity for new technological choices (and therefore also social, economic, local development, spatial organisation, etc.) to emerge. It could prove difficult to break away from past choices which are embedded not only in current technical solutions (mechanisation, fertilisers, pesticides, genetic engineering, etc.) but also in cognitive systems (knowledge and know-how, representations of nature, pollution, landscapes, etc.), in the values of the main actors involved (farmers as well as the services and administrations with which they deal), and in the prevailing techno-scientific reasoning and priorities given to other sectors of economic activity. In other words, like other business sectors, agriculture could find itself caught in a trap of technical rationalisation, in a sort of lock-in that will make alternative options difficult to design and implement. However the crucial role of agriculture for humanity as a whole does not allow for procrastination and for ignoring the risks involved in allowing current trends to continue.

Regulations for Trade and Sustainable Agriculture

One of the main findings of the Agrimonde foresight study is probably the necessity for accelerated development of international trade in food and agricultural products over the next few decades. Direct projections of the volumes of international trade involved in the various scenarios considered cannot be made in calculations of resource-use balances. Robust conclusions on this point can nevertheless be drawn from available figures, i.e. net trade balances (exports-imports) required to ensure that the needs of the world's large regions are met. We do so by adding up the regional deficits—essentially in terms of plant calories, according to the method of Variant 1 in which countries import animal feed rather than the animal products they lack (Chap. 8)—and interpreting the sum as an aggregated indicator of food trade. This is however a default estimate of the imports needed for regions with a net deficit since it fails to take into account trade among countries within the same region. Nor does it include the fact that certain countries can be net exporters of plant products and net importers of animal products, or vice-versa.

Despite these limitations, the calculation of this indicator for 2050 highlights a steep growth of food deficits in certain regions, in both scenarios, between the reference year (2003) and the two future situations foreseen. This total deficit of food products represents 9,933 Gkcal/day in Agrimonde 1 and 5,126 Gkcal/day in Agrimonde GO, whereas in 2003 the total net deficit was 1,307 (plant and animal food products). Even though, for reasons pertaining to methodology, the latter figure is not directly comparable to the total deficits of the 2050 scenarios (the simulations for 2050 assume, in Variant 1, a preference for importing animal feed, which was not entirely verified in the past), this comparison leads to extremely pronounced differences between 2003 and 2050, with a growing deficit of respectively 660 % and 292 % for the Agrimonde 1 and Agrimonde GO scenarios. Moreover, the total deficit of the regions with deficits in Agrimonde 1 is far greater than that envisaged in Agrimonde GO. The latter scenario is nevertheless supposed to represent an open world based on growing international trade, whereas the Agrimonde 1 scenario is more concerned with long-term sustainable development.

In total, the Agrimonde 1 scenario implies a steep increase in inter-regional and therefore international trade. Yet this scenario was built on a normative basis as regards the estimated food needs (a mean of 3,000 kcal/cap/day in all regions of the world) and environmental protection reflected in the choice of assumptions on surface areas and yields, discussed in Chaps. 6 and 7. *A priori*, these two choices seem to be consistent with the assumption of greater autonomy with regard to international markets. It is of course necessary to question the likelihood of these assumptions. The low variants of yields in the Agrimonde 1 scenario, in regions with deficits, are relatively weak, as noted above. They are far below a continuation of past trends, whereas assumptions of consumption, especially of animal calories in SSA, are high. For this region, a more optimistic assumption of increasing yields could have reduced the calorie deficit in 2050. In contrast, the deficit foreseen for ASIA and MENA seems unfortunately to be less debatable, given the region's low potential of cultivable land and their growth trends.

The result showing the necessity for international trade to increase seems robust, as is evidenced in numerous scenarios exploring a variety of assumptions on resources and uses of food biomass. In addition to the two Agrimonde scenarios, Michel Griffon's Doubly Green Revolution, more optimistic in terms of possible sustainable technical progress in the regions with shortages (ASIA, SSA and MENA), and other foresight and projection studies[11], conclude with the necessity for an increase in trade.

This growth of international trade unquestionably has significant implications as regards desirable regulation of commercial trade. Thus, a strong dose of protectionism, especially in regions with shortages, would not be advisable. Too much protection could impede the necessary growth of imports into such areas. Conversely, the assumptions of the Agrimonde 1 scenario, with regard to surface areas and yields in particular, imply economic viability of local agriculture, especially that based on many small semi-subsistence farms. However this economic viability could be undermined by competition from massive imports at cut prices, especially when exports are subsidised, as they were on a large scale by developed countries in the past. We therefore reach a conclusion that may seem surprising. We see that many poor countries, in regions with shortages, would have gained from a positive conclusion to the Doha Round, on the basis of the main measures that were "on the table" at the July 2008 WTO ministerial conference in Geneva (elimination of subsidies on exports, better access to the markets of developed countries, maintenance of the level of agricultural protection for the poorest countries, yet with an erosion of preferences for countries benefiting from privileged access to certain markets, such as the ACP Group of States in the European market[12]). The conference failed however, and analysing the causes of this failure falls outside the scope of the Agrimonde project. Most importantly, we need to emphasise the necessity for increased imports into regions with shortages, in both Agrimonde scenarios, and the danger of any ideological dogmatism as regards the regulation of international trade.

International trade will moreover have to be regulated in coherence with international action concerning the environment and, above all, the struggle against global warming, for agriculture is going to be heavily affected by climate change, especially in tropical countries. Recent negotiations on the subject, notably at the Copenhagen Conference in December 2009, illustrated the difficulties of including agriculture in programmes to cut greenhouse gas emissions and of adapting to the impacts of climate change. Yet, as the African countries vehemently affirmed, no global agreement is possible if it fails to explicitly include agriculture. Trade regulations are negotiated at the WTO, which has no authority when it comes to climate change. It is however clear, and generally acknowledged, that WTO rules will have to be aligned with any future agreement on climate change. Reaching an agreement on these issues within a reasonable timeframe is currently a formidable challenge for world governance.

[11] See for example the most recent results of the IMPACT model of the IFPRI (International Food Policy Research Institute), as published in the IAASTD Report (IAASTD 2009).

[12] The African, Caribbean and Pacific countries that signed the Lomé Convention and the Cotonou Agreement.

Chapter 12
Conclusion

Bernard Hubert, Patrick Caron and Hervé Guyomard

The Agrimonde foresight study enabled us to address questions that had not yet been formulated as such in mainstream scientific debates. Not that the facts reported or the assumptions put forward were entirely unpredictable or unknown; but they prompted the experts, followed by the various audiences to whom the results were presented, to consider the actions to be undertaken from a different point of view, primarily in relation to the type of future wished for—or not—in the decades to come. The value of baseline scenarios is not that they predict a sure future—which no one can actually anticipate as far ahead as 2050—but that they define the narrow path of possible futures that are likely to unfold if we fail to broaden our options now. Alternative scenario(s) can thus be mobilised to identify the objectives that we need to set when broadening the options explored. This has been the role of the Agrimonde 1 scenario, in contrast to the baseline scenario provided by Agrimonde GO.

We conclude this presentation of Agrimonde by very briefly reviewing the three main immediate challenges regarding food and agricultural systems. The idea is to be able to control our future better and for that purpose—in those areas where it seems necessary, for research institutions—to investigate the range of political and societal options. We then examine the role of foresight studies as back-up to agronomic research programming, in order to identify, in fine, some directions for future scenario-building in our organisations.

B. Hubert (✉)
Agropolis international, 34394 Montpellier Cedex 5, France
e-mail: bernard.hubert@avignon.inra.fr

P. Caron
CIRAD, avenue Agropolis, TA 179/04, 34398 Montpellier Cedex 5, France
e-mail: patrick.caron@cirad.fr

H. Guyomard
INRA, 147 rue de l'Université, 75338 Paris Cedex 7, France
e-mail: herve.Guyomard@rennes.inra.fr

S. Paillard et al. (eds.), *Agrimonde – Scenarios and Challenges for Feeding the World in 2050,* 197
DOI 10.1007/978-94-017-8745-1_12, © Éditions Quæ, 2014

Agrimonde in Relation to Three Challenges

Apart from figures and storylines, the comparison of the two scenarios enabled us to identify cross-cutting issues and three main challenges: (1) food consumption patterns; (2) technological and organisational options; and (3) trade in agricultural products and the regulation of that trade. For each of them, the comparison of the two scenarios yielded robust results that none of the experts of the panel had really anticipated.

Food Consumption Patterns: A Variable to Integrate before Production-Related Choices are Made!

As a rupture scenario, Agrimonde 1 is based on a convergence, at the level of the world's major regions, of apparent average daily per capita food consumption ('availabilities') on 3,000 kcal, 2,500 of which are from plants. This single value however corresponds to contrasting trends in the different areas. In sub-Saharan Africa, economic development and especially agricultural development will have to be strong enough to result in higher incomes, growth, and a diversification of food consumption. In contrast, in the developed countries of the OECD, food availabilities and the proportion of animal products in diets will have to decline, despite per capita income growth. Are these two major shifts imaginable? This question paves the way for three lines of thought:

- Food consumption trends can be broken down into three phases, according to income growth: a first phase of quantitative increases of consumption of all foods; a second phase of changing consumption, with more animal products, carbohydrates and saturated fats; and, finally, a third phase of stability in the composition of macronutrients. We can legitimately assume that most households in developed countries and in several emerging countries will be in the third phase of the cycle in 2050. Moreover, at that stage people over the age of 60 will account for over 20 % of the world's population (compared to 10 % today). As calorie needs decline in old-age, the aging of the world's population could also contribute to making the reduction of average food intake in rich countries plausible.
- The decrease in daily per capita food availability to 3,000 kcal should not be seen as a proportional reduction of the quantities ingested. It could also be due to the intake of food that is richer in fibres and micronutrients than in calories, such as fruit and vegetables, and to the reduction of loss and waste. At the retail and final consumption stage waste is considerable: approximately 800 kcal per capita per day in the developed countries. Consumption behaviours aimed at reducing such waste are still marginal but may gain in currency under the impulse of public policies. The need has also been highlighted to analyse, integrate and reduce the energy costs of the conservation, transport and distribution of food, something that has not yet been achieved.

- The Agrimonde 1 scenario assumes that the public regulator has taken the necessary steps to combat over-nourishment and related diseases. This is a major challenge. Too few measures have been taken and most are based on education and consumer information. Although necessary, such measures are however not sufficient. They should be part of more ambitious public health policies that make use of all available tools, based on an in-depth understanding of dietary patterns and their relationship with health.

Clearly, agronomic research cannot ignore issues of nutrition and diet. They are at the heart of the public health problems concerning all of our societies and particularly those in a phase of economic transition. Nutritional issues moreover raise questions on the production offer in all its dimensions. As this book has shown, the ways of addressing the problem of food security on a global scale and the difficulties encountered in the process will differ, depending on the food and nutritional options taken.

Technological and Organisational Choices in Agricultural Production: Options for Ecological Intensification

In the narrow sense of the term, the ecological intensification of agricultural systems at play in the Agrimonde 1 scenario can be defined as a type of agronomy that exploits ecological processes as much as possible. This new agronomy will require less fossil fuel. It will take advantage of the soil's ability to mobilise organic matter, for example by associating or sequencing certain crops and using new tillage techniques. Pests will be controlled by means of integrated pest management and production methods based on biological control organisms, crop rotation, longer rotation periods, or the organisation of fields. Crops will be more disease-resistant because a wider range of species and varieties will be used. Far from the backward-looking image sometimes associated with it, ecological intensification will take advantage of scientific and technological progress, for instance via the use of biotechnologies, remote sensing or marker-assisted selection. It will simultaneously exploit traditional knowledge and know-how.

Ecological intensification cannot be reduced to this technical dimension only. It must also be conceived of as a social, economic, political, spatial, etc. mode of organisation. To illustrate the implications of the spatial dimension analysed here in terms of the organisation of agricultural and natural areas, two models can be distinguished:

- a 'segregationist' model that distinguishes areas which are cultivated from those which, for environmental protection, are not. This model requires sound innovations to limit environmental damage in productive areas. However it is primarily in non-productive areas that global environmental problems and their resolution are treated, leading in the most extreme cases to the "sanctuarisation" of these areas,

- an 'integrationist' model which, in the same territory, associates different types of productive system and makes the territory a mosaic of ecosystems. These are simultaneously a source of commodities and of ecological services such as soil conservation, optimised water management, carbon storage, and protection of biodiversity. In this model, crop farming, livestock breeding, forestry, etc. coexist in the same territory and sometimes on the same farm. For instance, in Latin America forests are no longer cleared or protected; they are used for agroforestry in various ways suited to the sub-regional context. Likewise, in Middle East—North Africa and sub-Saharan Africa, grazing lands are both a source of animal feed and a key element in biodiversity protection. Farms which may not be efficient in terms of classical techno-economic performance play an essential environmental and social role, in which the concept of multifunctionality is fully meaningful. For this reason they can legitimately benefit from public support to ensure their survival, where necessary.

The challenge is to develop a range of technical options which are applicable in diverse situations and are often complementary to one another in terms of the types of resources, equipment, knowledge and know-how mobilised. Above all, these options must be oriented towards a more sustainable exploitation of resources by taking into account the way in which they are produced by agriculture itself on various spatial and temporal scales. For this purpose, one of the priorities is to define and agree on performance criteria that will make it possible to evaluate the effectiveness and efficiency of these renewed production systems, from the point of view of their intrinsic productivity, their effects on environmental parameters, and their economic viability and social sustainability with regard to the organisation and harshness of the tasks to accomplish.

International Trade in Agricultural and Agri-Food Products: A Necessity that Implies their Security

The picture in 2050 is the same in both scenarios: a world divided into two or three regions—OECD, Latin America and the former Soviet Union—which produce a surplus of food calories and make up for shortages in the other three regions—Asia, Middle East—North Africa, and sub-Saharan Africa. As, in both scenarios, the increase in agricultural production in the latter three areas is not sufficient to meet domestic needs, these countries have to import more food. Whereas optimistic assumptions of increasing yields could have been foreseen in sub-Saharan Africa given the poor yields today and the unexploited agricultural land in this region, the shortage foreseen for the regions of Asia and Middle East—North Africa in 2050 seems to be a fairly robust result, given the poor potential for extending cultivated areas and increasing yields.

This conclusion raises the question of how to secure international trade, in two respects. First, such trade has to make up for the deficit of domestic production in countries that are net importers of food calories, and thus configure an international environment that is as sure as possible. Secondly, it has to allow for increases in food production, requiring a stable international context from an economic, social and political point of view. Several approaches can be taken to secure increasing international trade. Far from all dogmatism and without claiming to provide a solution, the Agrimonde scenario-building exercise underscores the urgent need to conceive of future food and agricultural systems by integrating the international dimension and, more specifically, the question of global trade in agricultural and agri-food products, its security, its stability, and relations with international environmental and social regulations.

Agrimonde, a Tool for Collective Reflection and Research Programming

In the first two scenarios (Agrimonde 1 and Agrimonde GO) built and analysed in the Agrimonde study, it seems possible to feed the planet in 2050. Moreover, this could be achieved through sustainable development of the world's food and agricultural systems. Several conditions would however have to be met, and, as most of these conditions represent shifts in current trends, this would be no easy task.

The Agrimonde 1 scenario outlines a path—among others—for the development of sustainable food and agricultural systems in the world. Like any scenario, it is based on assumptions which need to be satisfied if it is to be feasible. In this case we assume that the developing world's economic growth will basically be agricultural and rural, that demographic growth and international migration will remain close to the current average forecasts, and that climate change and its effects will be relatively moderate. Agrimonde 1 is also based on three main changes: the reduction of excessive food consumption and waste; the development of an ecologically intensive agronomy; and the security of international trade in agricultural and agri-food products. These three interdependent trends have to be envisaged in multiple dimensions: technical, social, economic and political. They require significant investments, a substantial research and research-development effort, and new public policies at international and national level.

The food question is complex, both per se and in relation to the environment, energy, health, and the economic development of rural areas and developing countries. This complexity should not be a source of concern, causing opposition to change and thus resulting in the Business-As-Usual scenario—which we know is not viable for humanity. The foresight approach requires us to confront the situation and to anticipate the expected future challenges.

In the agricultural and agri-food domain, as in others, initiatives involving scientists on a global scale have proliferated over the past few years. These initiatives concern the organisation and regulation of trade by the World Trade Organization (WTO), the review and analysis of available knowledge within the framework of the IAASTD[1], studies undertaken at the initiative of the World Bank[2], the organisation of the global research system within the framework of the Global Forum on Agricultural Research[3] and of the current reform of the CGIAR[4], and the present creation of the GPAFS[5]. Unlike climate change, these studies—advice, assessment, foresight, research—are a continuation of the issues and political debates that preceded them. They are the basis for 'rational' scientific thinking, 'commonly accepted by the experts', and for the international standards that it spawns. The implications are considerable because this thinking steers the choices of international organisations and governments. It weighs on negotiations in the fields of agriculture, the environment, development and trade, as well as on agronomic research. The quality of analytical studies can cause representations to evolve and new spaces to be opened for action, especially public action, including internationally.

The first debates, both in France and on the international scene, on the Agrimonde GO and Agrimonde 1 scenarios, their results and their implications have already enabled us to enrich our initial analysis, to make new assumptions, and to consider new alternative futures. Additionally, they have shown that studies seemingly as different as the IAASTD, the projections of the International Food Policy Research Institute, of the FAO and of the Millennium Institute, and the foresights of the European Union's Standing Committee on Agricultural Research (SCAR), of the International Centre for Advanced Mediterranean Agronomic Studies (CIHEAM) or of the UK government all point to the fact that food security is above all a problem of access to food by the poorest populations. It is not only a question of production. The situation of the poorest, especially in rural areas, must be seen as a priority for both research and action, especially since they are going to be the first to be affected by the deterioration of the environment and climate change. There is likewise convergence on the necessity to undertake research on energy consumption in agriculture, on relations between urban and rural areas, on situations of demographic transition and their consequences on the organisation of labour and diet, on the reduction of loss and waste, etc. Finally, all these organisations highlight the need for a broader perspective on the agricultural activities in changing rural worlds, in all regions of the world and in all economic dynamics, including intersectoral approaches.

---

[1] International Assessment of Agricultural Knowledge, Science and Technology for Development. The reports stemming from this global collective assessment were published in 2009 by Island Press; they are available at: http://www.agassessment.org.

[2] In 2008, the annual report on the state of development in the world, commissioned by the World Bank (World Bank 2008), was devoted to agriculture for the first time in 27 years.

[3] http://www.egfar.org

[4] Consultative Group on International Agricultural Research which coordinates the action of 15 international agronomic research centres created since the 1970s (http://www.cgiar.org).

[5] Global Partnership for Agriculture and Food Security currently being created following the June 2008 FAO summit on food, convened in reaction to the 'food crisis' at the beginning of that year, the summit organised in early 2009 in Madrid by the United Nations (http://www.ransa2009.org), and the Heads of State summit held in Rome in November 2009.

INRA and CIRAD have the ambition, if not the duty, to continue and to intensify the effort in this direction. It is today's decisions that will determine the trajectories which will enable tomorrow's world to adequately meet all its citizens' needs in food, energy and biomaterials, to reduce poverty and inequalities, and to curb the deterioration of environmental goods and services. In a world of rare resources, the rarest of all may be time.

Clearly, it will be useful, if not necessary, to learn to effectively bring together two worlds that have different underlying values and paradigms: the world of foresight and that of research. By helping research to identify the right questions on the future, but which need to be addressed today, foresight can help us avoid intellectual dead-ends focused on the production of evidence-based arguments which are supposed to contribute to debates at the highest level. An example is the call for genetic improvement as the only way of reducing famine, by increasing productivity. While this solution should of course not be excluded, it is clearly insufficient. Foresight can also help researchers and decision-makers not to fall into the trap of doing nothing, on the pretext that we do not know everything about such complex issues. It teaches us, on the contrary, to address problems and to anticipate emerging challenges for the future.

Faced with the growing complexity of challenges and knowledge, there is currently virtually no international organisation, country or NGO that does not intervene in one way or another in research and debates on the future of global agricultural and food production, their interactions with the objective of sustainable development, and their implications for international relations and public policies. It is true that the profusion of information, data and results, as well as the consequent overall illegibility, are not an encouragement to examine the underlying hypotheses—whether scientific or ideological—or to give the space they warrant to analyses that do not correspond to the international *doxa*. Yet we must aim to enhance and expand these studies, so that we can illuminate the future options towards which we should be tending. This is what INRA and CIRAD have started to do with this foresight study, and what they will continue to do, in the framework of a broader partnership. The first results reported and commented on in this book show how important it is to create frameworks for defining the relevant questions to be addressed. That is what Agrimonde is all about.

Appendix

Appendix 1

Lists of Countries and FAO-SUA Product Lines Used in Agribiom Comparison of Observed and Simulated Animal Food Productions (Table A1.1 and A1.2; Fig. A1.1)

Appendix 2

Review of the World Food Economy (Fig. A2.1, A2.2, A2.3, A2.4, A2.5, A2.6, A2.7, A2.8 and A2.9)

Appendix 3

Food Availability and Loss of Food Calories in the Agrimonde GO Scenario Definition of Loss and Waste of Food Calories

Establishing Food Availability Assumptions in the Agrimonde GO Scenario

Total regional and global food availabilities are provided in the Millennium Ecosystem Assessment (MA) report, however this document provides no breakdown by product (MA 2005b).

S. Paillard et al. (eds.), *Agrimonde – Scenarios and Challenges for Feeding the World in 2050*,
DOI 10.1007/978-94-017-8745-1,

Table A1.1 Countries selection and their allocation in MA regions

Country		Statistical period	Region allocation	
English name	French name		MA region	Region fullname
Albania	Albanie	1961–2003	OECD	OECD 1990
Algeria	Algérie	1961–2003	MENA	Middle East and North Africa
Angola	Angola	1961–2003	SSA	Sub-Saharan Africa
Argentina	Argentine	1961–2003	LAM	Latin America
Armenia	Arménie	1992–2003	FSU	Former Soviet Union
Australia	Australie	1961–2003	OECD	OECD 1990
Austria	Autriche	1961–2003	OECD	OECD 1990
Azerbaijan, Republic of	Azerbaïdjan, République de	1992–2003	FSU	Former Soviet Union
Bangladesh	Bangladesh	1961–2003	ASIA	Asia—Asie
Belarus	Bélarus	1992–2003	FSU	Former Soviet Union
Belgium-Luxembourg	Belgique-Luxembourg	1961–2003	OECD	OECD 1990
Benin	Bénin	1961–2003	SSA	Sub-Saharan Africa
Bolivia	Bolivie	1961–2003	LAM	Latin America
Bosnia and Herzegovina	Bosnie-Herzégovine	1992–2003	OECD	OECD 1990
Botswana	Botswana	1961–2003	SSA	Sub-Saharan Africa
Brazil	Brésil	1961–2003	LAM	Latin America
Bulgaria	Bulgarie	1961–2003	OECD	OECD 1990
BurkinaFaso	Burkina Faso	1961–2003	SSA	Sub-Saharan Africa
Burundi	Burundi	1961–2003	SSA	Sub-Saharan Africa
Cambodia	Cambodge	1961–2003	ASIA	Asia
Cameroon	Cameroun	1961–2003	SSA	Sub-Saharan Africa
Canada	Canada	1961–2003	OECD	OECD 1990
Central African Republic	Centrafricaine, République	1961–2003	SSA	Sub-Saharan Africa
Chad Tchad		1961–2003	SSA	Sub-Saharan Africa
Chile	Chili	1961–2003	LAM	Latin America
China	Chine	1961–2003	ASIA	Asia
Colombia	Colombie	1961–2003	LAM	Latin America
Congo, Democratic Republic	of Congo, République démocratique du	1961–2003	SSA	Sub-Saharan Africa
Congo, Republic of	Congo, République du	1961–2003	SSA	Sub-Saharan Africa
Costa Rica	Costa Rica	1961–2003	LAM	Latin America
Côte d'Ivoire	Côte d'Ivoire	1961–2003	SSA	Sub-Saharan Africa
Croatia	Croatie	1992–2003	OECD	OECD 1990
Cuba	Cuba	1961–2003	LAM	Latin America
Czech Republic	Tchèque, République	1993–2003	OECD	OECD 1990
Czechoslovakia	Tchécoslovaquie	1961–1992	OECD	OECD 1990
Denmark	Danemark	1961–2003	OECD	OECD 1990
Dominican Republic	Dominicaine, République	1961–2003	LAM	Latin America

Table A1.1 (continued)

Country		Statistical period	Region allocation	
English name	French name		MA region	Region fullname
Ecuador	Équateur	1961–2003	LAM	Latin America
Egypt	Égypte	1961–2003	MENA	Middle East and North Africa
El Salvador	El Salvador	1961–2003	LAM	Latin America
Eritrea	Érythrée	1993–2003	SSA	Sub-Saharan Africa
Estonia	Estonie	1992–2003	OECD	OECD 1990
Ethiopia	Éthiopie	1993–2003	SSA	Sub-Saharan Africa
Ethiopia PDR	Éthiopie RDP	1961–1992	SSA	Sub-Saharan Africa
Finland	Finlande	1961–2003	OECD	OECD 1990
France	France	1961–2003	OECD	OECD 1990
Gabon	Gabon	1961–2003	SSA	Sub-Saharan Africa
Gambia	Gambie	1961–2003	SSA	Sub-Saharan Africa
Georgia	Géorgie	1992–2003	FSU	Former Soviet Union
Germany	Allemagne	1961–2003	OECD	OECD 1990
Ghana	Ghana	1961–2003	SSA	Sub-Saharan Africa
Greece	Grèce	1961–2003	OECD	OECD 1990
Guatemala	Guatemala	1961–2003	LAM	Latin America
Guinea	Guinée	1961–2003	SSA	Sub-Saharan Africa
Guinea-Bissau	Guinée-Bissau	1961–2003	SSA	Sub-Saharan Africa
Guyana	Guyana	1961–2003	LAM	Latin America
Haiti	Haïti	1961–2003	LAM	Latin America
Honduras	Honduras	1961–2003	LAM	Latin America
Hungary	Hongrie	1961–2003	OECD	OECD 1990
Iceland	Islande	1961–2003	OECD	OECD 1990
India	Inde	1961–2003	ASIA	Asia
Indonesia	Indonésie	1961–2003	ASIA	Asia
Iran, Islamic Republic of	Iran, République islamique d'	1961–2003	MENA	Middle East and North Africa
Ireland	Irlande	1961–2003	OECD	OECD 1990
Israel	Israël	1961–2003	MENA	Middle East and North Africa
Italy	Italie	1961–2003	OECD	OECD 1990
Jamaica	Jamaïque	1961–2003	LAM	Latin America
Japan	Japon	1961–2003	OECD	OECD 1990
Jordan	Jordanie	1961–2003	MENA	Middle East and North Africa
Kazakhstan	Kazakhstan	1992–2003	FSU	Former Soviet Union
Kenya	Kenya	1961–2003	SSA	Sub-Saharan Africa
Korea, Democratic People's Republic	Corée, République populaire démocratique	1961–2003	ASIA	Asia
Korea, Republic of	Corée, République de	1961–2003	ASIA	Asia
Kuwait	Koweït	1961–2003	MENA	Middle East and North Africa
Kyrgyzstan	Kirghizistan	1992–2003	FSU	Former Soviet Union
Laos	Laos	1961–2003	ASIA	Asia

Table A1.1 (continued)

Country		Statistical period	Region allocation	
English name	French name		MA region	Region fullname
Latvia	Lettonie	1992–2003	OECD	OECD 1990
Lebanon	Liban	1961–2003	MENA	Middle East and North Africa
Lesotho	Lesotho	1961–2003	SSA	Sub-Saharan Africa
Liberia	Libéria	1961–2003	SSA	Sub-Saharan Africa
Libyan Arab Jamahiriya	Libyen, Jamahiriya arabe	1961–2003	MENA	Middle East and North Africa
Lithuania	Lituanie	1992–2003	OECD	OECD 1990
Macedonia, The Fmr Yug Republic	Macédoine, l'ex-République Yougoslavie	1992–2003	OECD	OECD 1990
Madagascar	Madagascar	1961–2003	SSA	Sub-Saharan Africa
Malawi	Malawi	1961–2003	SSA	Sub-Saharan Africa
Malaysia	Malaisie	1961–2003	ASIA	Asia
Mali	Mali	1961–2003	SSA	Sub-Saharan Africa
Mauritania	Mauritanie	1961–2003	SSA	Sub-Saharan Africa
Mexico	Mexique	1961–2003	LAM	Latin America
Moldova, Republic of	Moldova, République de	1992–2003	FSU	Former Soviet Union
Mongolia	Mongolie	1961–2003	ASIA	Asia
Morocco	Maroc	1961–2003	MENA	Middle East and North Africa
Mozambique	Mozambique	1961–2003	SSA	Sub-Saharan Africa
Myanmar	Myanmar	1961–2003	ASIA	Asia
Namibia	Namibie	1961–2003	SSA	Sub-Saharan Africa
Nepal	Népal	1961–2003	ASIA	Asia
Netherlands	Pays-Bas	1961–2003	OECD	OECD 1990
New Zealand	Nouvelle-Zélande	1961–2003	OECD	OECD 1990
Nicaragua	Nicaragua	1961–2003	LAM	Latin America
Niger	Niger	1961–2003	SSA	Sub-Saharan Africa
Nigeria	Nigeria	1961–2003	SSA	Sub-Saharan Africa
Norway	Norvège	1961–2003	OECD	OECD 1990
Pakistan	Pakistan	1961–2003	ASIA	Asia
Panama	Panama	1961–2003	LAM	Latin America
Paraguay	Paraguay	1961–2003	LAM	Latin America
Peru	Pérou	1961–2003	LAM	Latin America
Philippines	Philippines	1961–2003	ASIA	Asia
Poland	Pologne	1961–2003	OECD	OECD 1990
Portugal	Portugal	1961–2003	OECD	OECD 1990
Romania	Roumanie	1961–2003	OECD	OECD 1990
Russian Federation	Fédération de Russie	1992–2003	FSU	Former Soviet Union
Rwanda	Rwanda	1961–2003	SSA	Sub-Saharan Africa
Saudi Arabia	Arabie saoudite	1961–2003	MENA	Middle East and North Africa
Senegal	Sénégal	1961–2003	SSA	Sub-Saharan Africa
Sierra Leone	Sierra Leone	1961–2003	SSA	Sub-Saharan Africa

Table A1.1 (continued)

Country		Statistical period	Region allocation	
English name	French name		MA region	Region fullname
Slovakia	Slovaquie	1993–2003	OECD	OECD 1990
Slovenia	Slovénie	1992–2003	OECD	OECD 1990
Solomon Islands	Salomon, Îles	1961–2003	ASIA	Asia
South Africa	Afrique du Sud	1961–2003	SSA	Sub-Saharan Africa
Spain	Espagne	1961–2003	OECD	OECD 1990
Sri Lanka	Sri Lanka	1961–2003	ASIA	Asia
Sudan	Soudan	1961–2003	SSA	Sub-Saharan Africa
Suriname	Suriname	1961–2003	LAM	Latin America
Swaziland	Swaziland	1961–2003	SSA	Sub-Saharan Africa
Sweden	Suède	1961–2003	OECD	OECD 1990
Switzerland	Suisse	1961–2003	OECD	OECD 1990
Syrian Arab Republic	Syrienne, République arabe	1961–2003	MENA	Middle East and North Africa
Tajikistan	Tadjikistan	1992–2003	FSU	Former Soviet Union
Tanzania, United Republic of	Tanzanie, République unie de	1961–2003	SSA	Sub-Saharan Africa
Thailand	Thaïlande	1961–2003	ASIA	Asia
Timor-Leste	Timor oriental	1961–2003	ASIA	Asia
Togo	Togo	1961–2003	SSA	Sub-Saharan Africa
Tunisia	Tunisie	1961–2003	MENA	Middle East and North Africa
Turkey	Turquie	1961–2003	MENA	Middle East and North Africa
Turkmenistan	Turkménistan	1992–2003	FSU	Former Soviet Union
Uganda	Ouganda	1961–2003	SSA	Sub-Saharan Africa
Ukraine	Ukraine	1992–2003	FSU	Former Soviet Union
United Arab Emirates	Émirats arabes unis	1961–2003	MENA	Middle East and North Africa
United Kingdom	Royaume-Uni	1961–2003	OECD	OECD 1990
United States of America	États-Unis d'Amérique	1961–2003	OECD	OECD 1990
Uruguay	Uruguay	1961–2003	LAM	Latin America
USSR (ex-)	URSS (ex-)	1961–1991	FSU	Former Soviet Union
Uzbekistan	Ouzbékistan	1992–2003	FSU	Former Soviet Union
Venezuela, Boliv Republic of	Venezuela, République boliv du	1961–2003	LAM	Latin America
Viet Nam	Viet Nam	1961–2003	ASIA	Asia
Yemen	Yémen	1961–2003	MENA	Middle East and North Africa
Yugoslavia SFR	Yougoslavie FRS	1961–1991	OECD	OECD 1990
Zambia	Zambie	1961–2003	SSA	Sub-Saharan Africa
Zimbabwe	Zimbabwe	1961–2003	SSA	Sub-Saharan Africa

Table A1.2 FAO-SUA product lines and their allocation into compartments of food biomass

Biomass		Allocation	
English name	French name	Compartment	Species of origin
Wheat	Blé	VEGE	Cere (cereal)
Rice (Milled Eq)	Riz (Eq blanchi)	VEGE	Cere
Ricebran Oil	Huile de son de riz	VEGE	Cere
Barley	Orge	VEGE	Cere
Maize	Maïs	VEGE	Cere
Maize Germ Oil	Huile de germe de maïs	VEGE	Cere
Rye	Seigle	VEGE	Cere
Oats	Avoine	VEGE	Cere
Millet	Millet	VEGE	Cere
Sorghum	Sorgho	VEGE	Cere
Cereals, Other	Céréales, Autres	VEGE	Cere
Brans	Sons	VEGE	Cere
Cassava	Manioc	VEGE	Root
Potatoes	Pommes de terre	VEGE	Root
Sweet Potatoes	Patate douce	VEGE	Root
Yams	Ignames	VEGE	Root
Roots, Other	Racines, autre	VEGE	Root
Sugar Cane	Canne à sucre	VEGE	Suga (sugar plant)
Sugar Beet	Betteraves à sucre	VEGE	Suga
Sugar, Non-Centrifugal	Sucre, non-centrifugé	VEGE	Suga
Sugar, Raw Equivalent	Sucre, éq. brut	VEGE	Suga
Molasses	Mélasse	VEGE	Suga
Beans	Haricots	VEGE	Puls (pulses)
Peas	Pois	VEGE	Puls
Pulses, Other	Légumineuses, autres	VEGE	Puls
Treenuts	Fruit coque	VEGE	Olea (oilseed)
Soyabeans	Soja (fèves)	VEGE	Olea
Groundnuts (Shelled Eq)	Arachide (décortiquées)	VEGE	Olea
Sunflowerseed	Tournesol (Graines)	VEGE	Olea
Rape & Mustardseed	Colza, moutarde (graines)	VEGE	Olea
Cottonseed	Coton (graines)	VEGE	Olea
Coconuts (Incl Copra)	Coco (inclus le coprah)	VEGE	Olea
Sesameseed	Sésame (graines)	VEGE	Olea
Palmkernels	Palme (amandes)	VEGE	Olea
Olives	Olives	VEGE	Olea
Oilcrops, Other	Plantes oléifères, autres	VEGE	Olea
Soyabean Oil	Huile de soja	VEGE	Olea
Groundnut Oil	Huile d'arachide	VEGE	Olea
Sunflowerseed Oil	Huile de tournesol	VEGE	Olea
Rape & Mustard Oil	Huile de colza, moutarde	VEGE	Olea
Cottonseed Oil	Huile graines de coton	VEGE	Olea
Sesameseed Oil	Huile de sésame	VEGE	Olea
Oilcrops Oil, Other	Huiles végétales, Autres	VEGE	Olea
Soyabean Cake	Tourteau de soja	VEGE	Olea
Groundnut Cake	Tourteau d'arachide	VEGE	Olea

Table A1.2 (continued)

Biomass		Allocation	
English name	French name	Compartment	Species of origin
Sunflowerseed Cake	Tourteau de tournesol	VEGE	Olea
Rape and Mustard Cake	Tourteau de colza et de moutarde	VEGE	Olea
Cottonseed Cake	Tourteau de coton	VEGE	Olea
Sesameseed Cake	Tourteau de sésame	VEGE	Olea
Oilseed Cakes, Other	Tourteau, autres	VEGE	Olea
Palmkernel Oil	Huile de palmistes	VEGE	Olea
Palm Oil	Huile de palme	VEGE	Olea
Coconut Oil	Huile de coco	VEGE	Olea
Olive Oil	Huile d'olive	VEGE	Olea
Palmkernel Cake	Tourteau de palmiste	VEGE	Olea
Copra Cake	Tourteau de coprah	VEGE	Olea
Tomatoes	Tomates	VEGE	Vege (vegetables)
Onions	Oignons	VEGE	Vege
Vegetables, Other	Légumes, autres	VEGE	Vege
Oranges, Mandarines	Oranges, mandarines	VEGE	Frui (fruit)
Lemons, Limes	Citrons, limes	VEGE	Frui
Grapefruit	Pamplemousse	VEGE	Frui
Citrus, Other	Agrumes, autres	VEGE	Frui
Bananas	Banane	VEGE	Frui
Plantains	Plantains	VEGE	Frui
Apples	Pommes	VEGE	Frui
Pineapples	Ananas	VEGE	Frui
Dates	Datte	VEGE	Frui
Grapes	Raisin	VEGE	Frui
Fruits, Other	Fruits, Autres	VEGE	Frui
Sweeteners, Other	Édulcorants, autres	VEGE	Othe (other)
Honey	Miel	VEGE	Othe
Pimento	Piments	VEGE	Othe
Spices, Other	Épices, autres	VEGE	Othe
Misc. Food	Divers alimentaire	VEGE	Othe
Coffee	Café	VEGE	Othe
Cocoa Beans	Fève de cacao	VEGE	Othe
Tea	Thé	VEGE	Othe
Pepper	Poivre	VEGE	Othe
Cloves	Clou	VEGE	Othe
Wine	Vin	VEGE	Alco (alcohol)
Beer	Bière	VEGE	Alco
Beverages, Fermented	Boissons fermentées	VEGE	Alco
Beverages, Alcoholic	Boissons alcoolisées	VEGE	Alco
Bovine Meat	Viande de bovins	RUMI	Rumi (ruminant)
Mutton & Goat Meat	Viande d'ovins, caprins	RUMI	Rumi
Meat, Other	Viande, autres	RUMI	Rumi
Offals, Edible	Abats comestible	RUMI	Rumi
Milk (Excl Butter)	Lait (Excl Beurre)	RUMI	Rumi
Butter, Ghee	Beurre, Ghee	RUMI	Rumi
Cream	Crème	RUMI	Rumi

Table A1.2 (continued)

Biomass		Allocation	
English name	French name	Compartment	Species of origin
Fats, Animals (Raw)	Graisses animales (Crue)	RUMI	Rumi
Meat Meal	Farines de viande	RUMI	Rumi
Pigmeat	Viande de porc	MONO	Pigs (pig)
Poultry Meat	Viande de volailles	MONO	Poul (poultry)
Eggs	Œufs	MONO	Poul
Freshwater Fish	Poissons d'eau douce	AQUA	Aqua (freshwaters species)
Fish, Body Oil	Huiles de poissons	MARI	Mari (marine species)
Fish, Liver Oil	Huiles de foie de poisson	MARI	Mari
Demersal Fish	Perciform	MARI	Mari
Pelagic Fish	Poissons pélagiques	MARI	Mari
Marine Fish, Other	Poissons marins, autres	MARI	Mari
Crustaceans	Crustacés	MARI	Mari
Cephalopods	Céphalopodes	MARI	Mari
Molluscs, Other	Mollusques, autres	MARI	Mari
Meat, Aquatic Mammals	Viande d'animaux aquatiques	MARI	Mari
Aquatic Animals, Other	Animaux Aquatiques, Autres	MARI	Mari
Fish Meal	Farines de poisson	MARI	Mari
Aquatic Plants	Plantes aquatiques	MARI	Mari

To quantify the calories available by origin (plant, grazing/monogastric animal, fresh water/marine) for each region in the Global Orchestration scenario of the MA, we extrapolated as follows.

Land Animal Products

Regional meat and cereal consumption trends in kilograms between 1997 and 2050 are provided in the MA report. These trends in the Global Orchestration scenario served as a basis to quantify the availability of animal calories in the Agrimonde GO scenario. By applying this trend coefficient to figures on the availability of food from land animals (meat, milk and dairy products, eggs, etc.), for 2000, we were able to extrapolate the availability of animal calories for each region in Agrimonde GO.

NB: Food availabilities in 2000 were calculated on the basis of FAOSTAT 1 data and are the same in Agrimonde 1 and Agrimonde GO (Chap. 2).

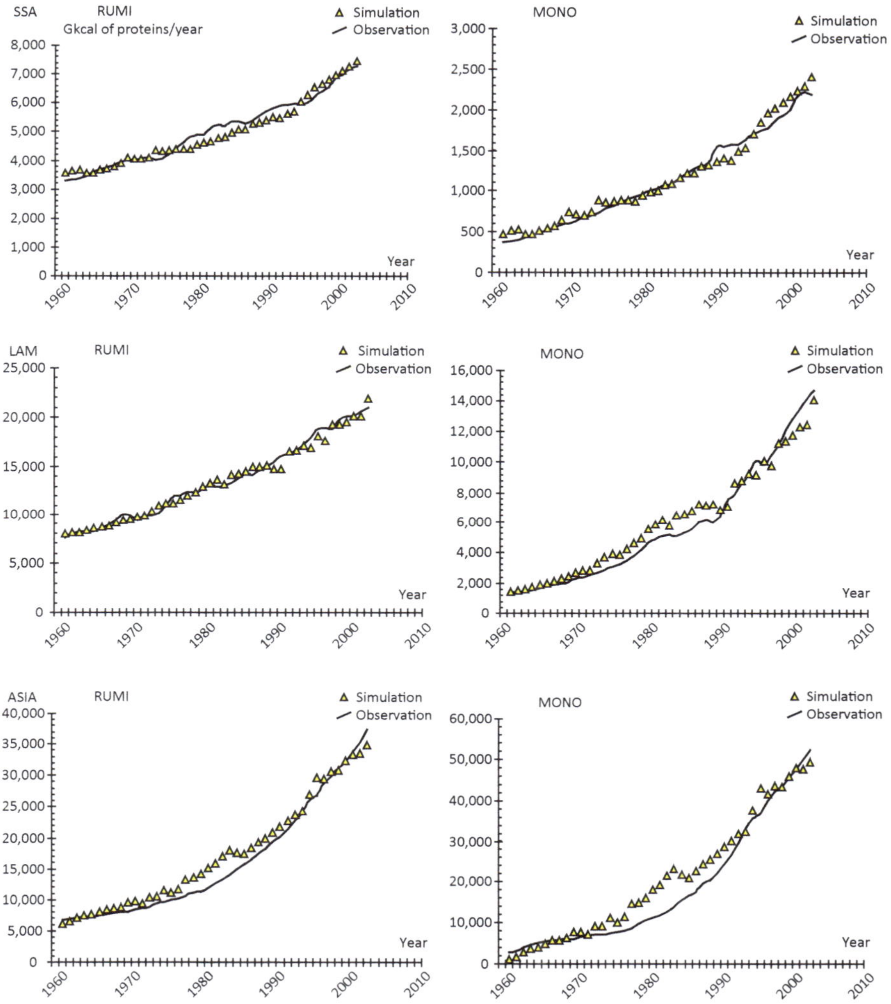

Fig. A1.1 Observed and simulated animal food productions (1961–2003). Comparison of 1961–2003 animal food productions observed and simulated (Gkcal of proteins), with regional production functions used for the Agrimonde foresight, in two categories of animal product: ruminants and large herbivore animals (*RUMI*, grazing animals), monogastric animals (*MONO*, non-grazing animals). Production estimates and production simulations by MA region, 1961–2003. (Source: B. Dorin & T. Le Cotty, based on FAO data)

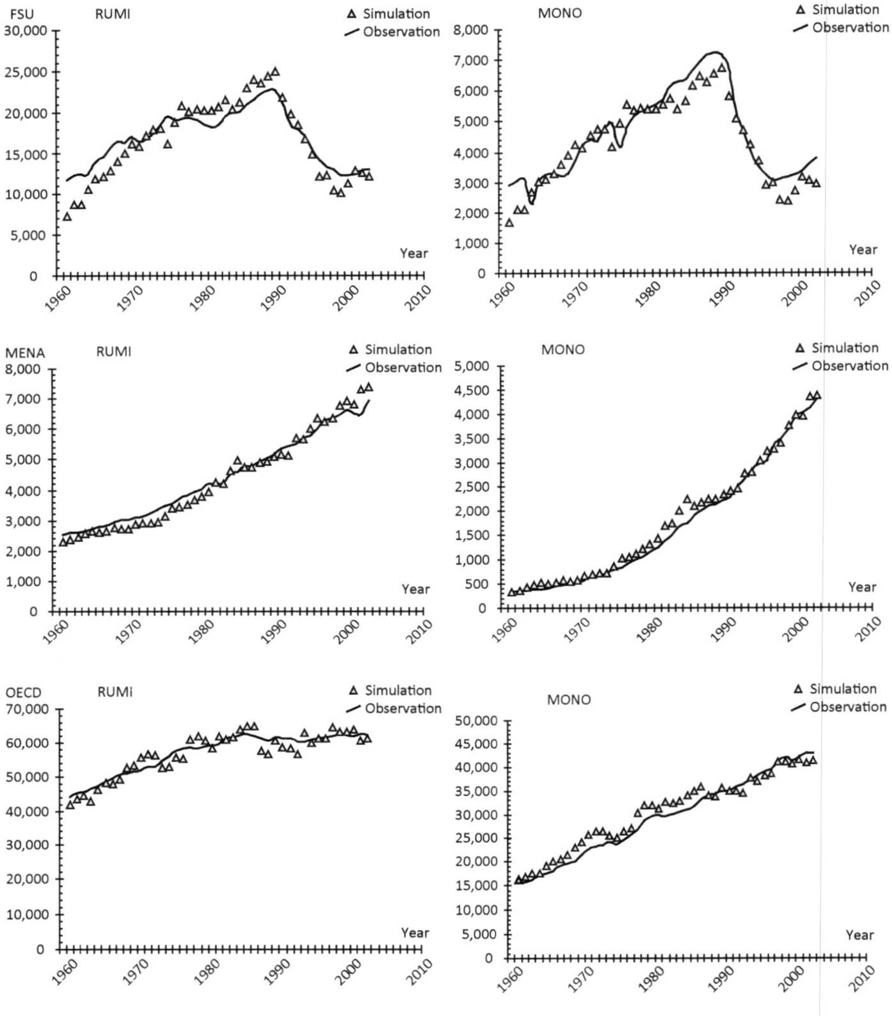

Fig. A1.1 (continued)

Aquatic Products

Qualitative data on fishing and aquaculture in the MA scenarios have been converted into annual growth rates of regional production. In so far as, for sake of simplicity, we assume in the Agrimonde scenarios that the apparent regional consumption of aquatic products is equal to the regional production, these annual production

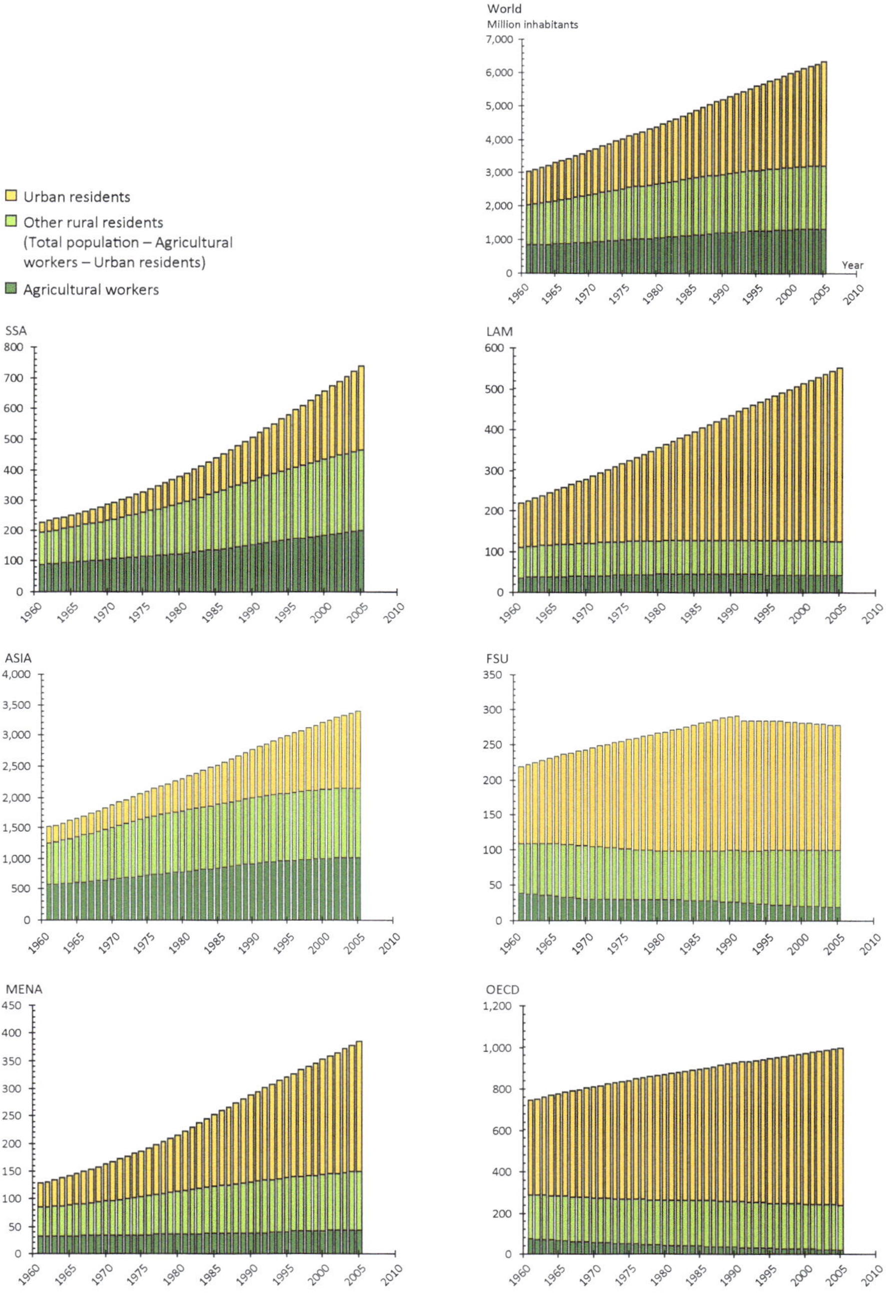

Fig. A2.1 Human populations (1961–2005). Human populations (million inhabitants) under 3 categories: World estimates (Agribiom countries) and by MA region. (Source: B. Dorin, computed from FAO data)

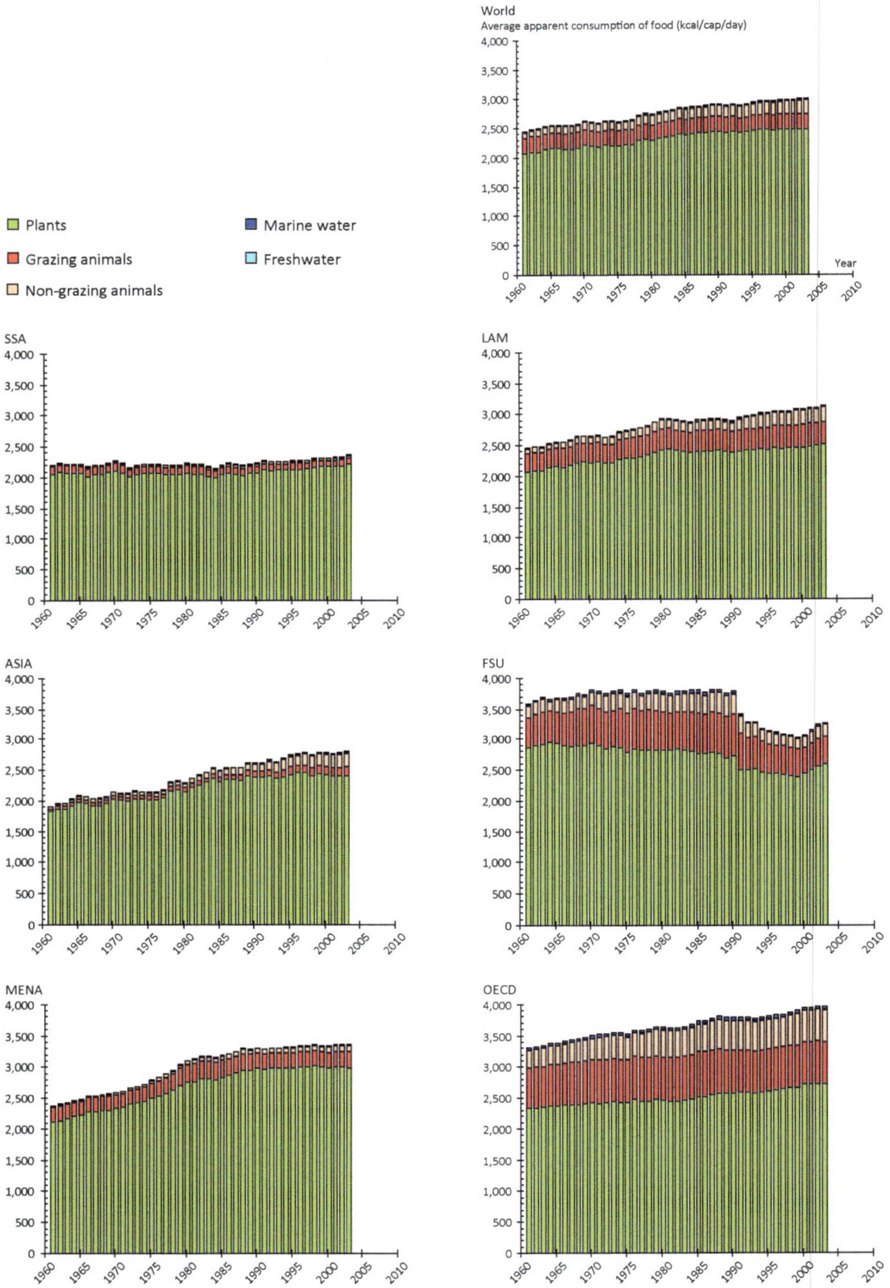

Fig. A2.2 Food availabilities (1961–2003). Average per capita apparent consumption of food (kcal/cap/day) under 5 categories of food origin: World estimates (Agribiom countries) and by MA region. (Source: B. Dorin, computed from FAO data)

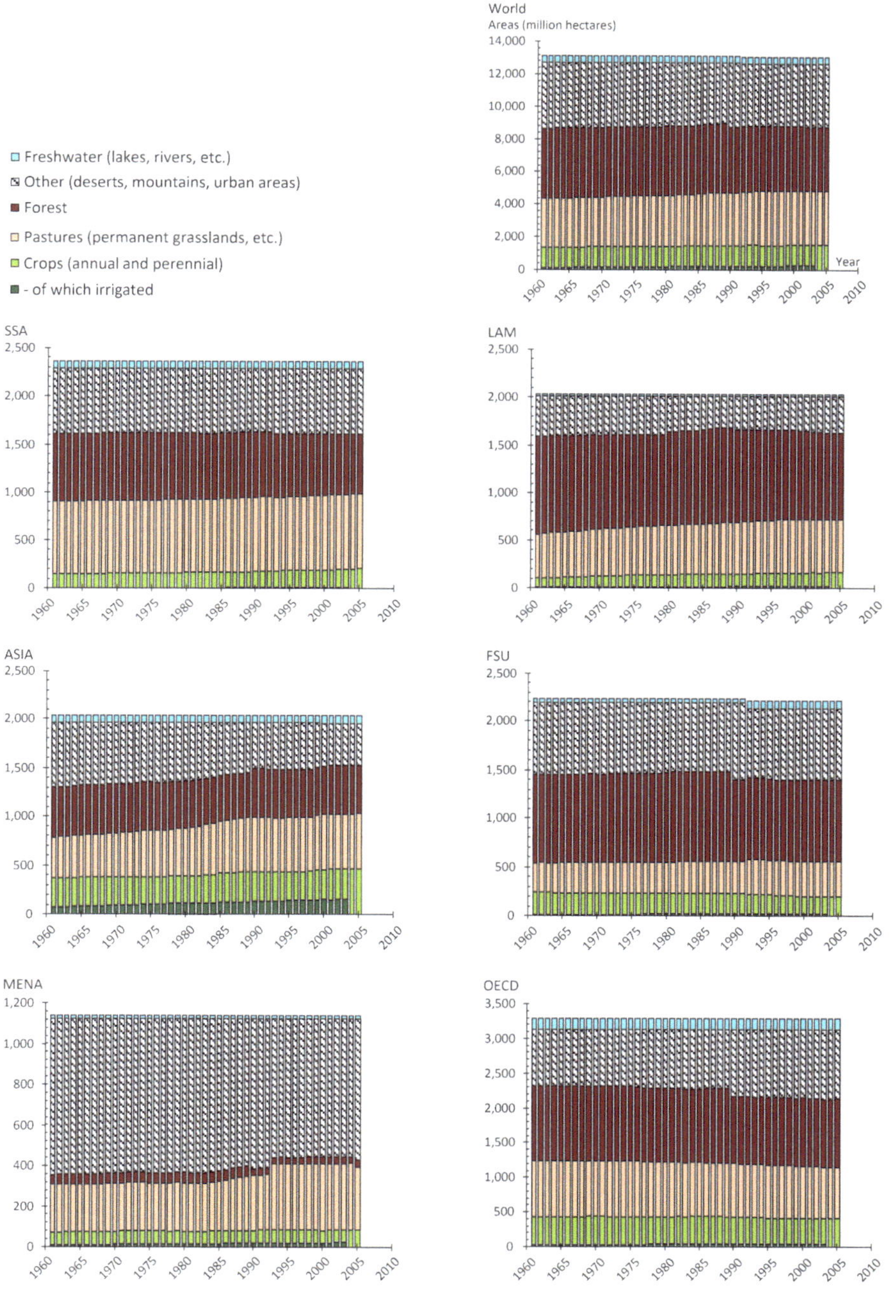

Fig. A2.3 Land uses (1961–2005). Land area (million hectares) under 5 categories of occupation: World estimates (Agribiom countries) and by MA region. (Soure: B. Dorin, computed from FAO data)

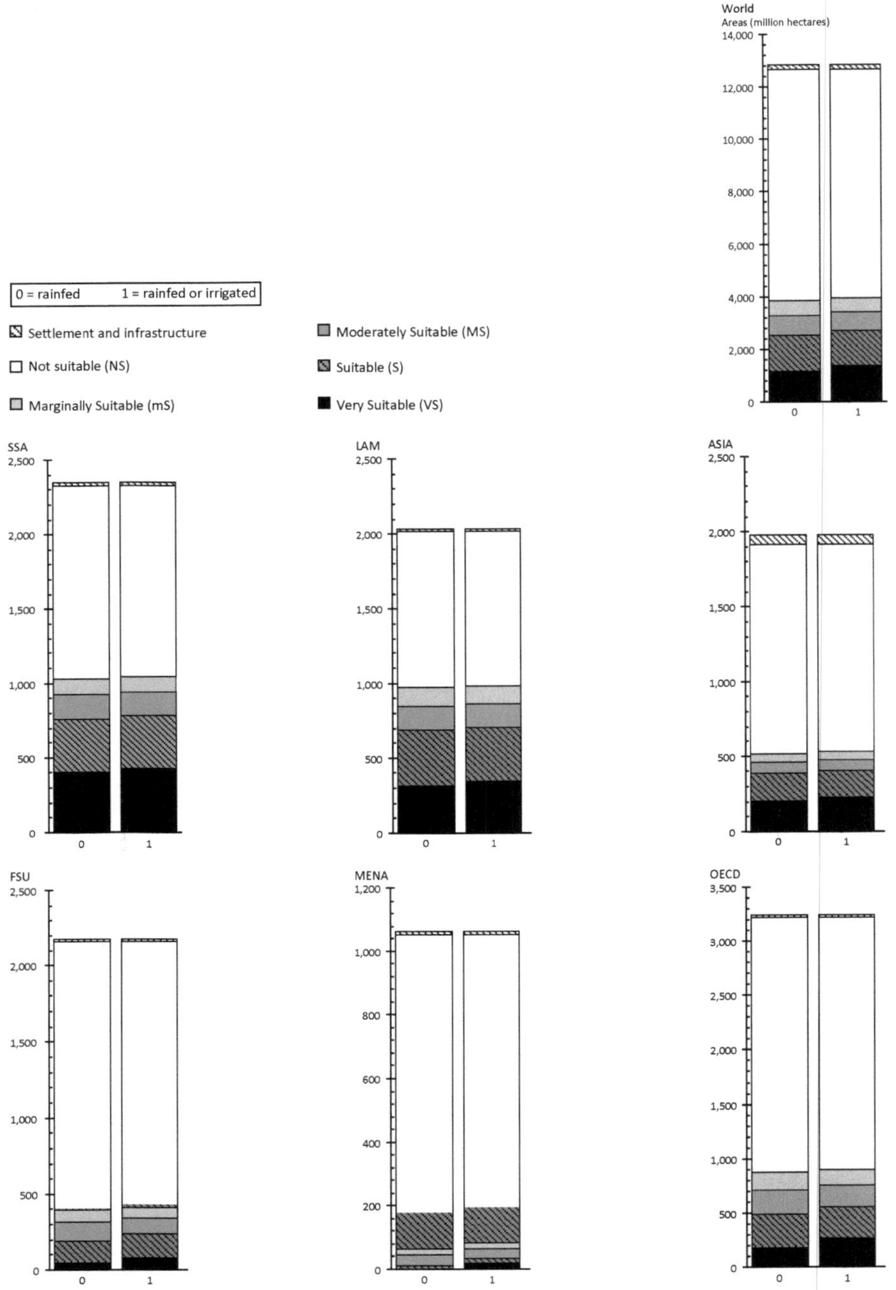

Fig. A2.4 Potential croplands (2000). Land area (million hectares) under 6 categories of potential croplands (crops: "all", input level: "mix"): World estimates (Agribiom countries) and by MA region. (Source: B. Dorin, computed from IIASA-FAO data (GAEZ, Fischer et al.))

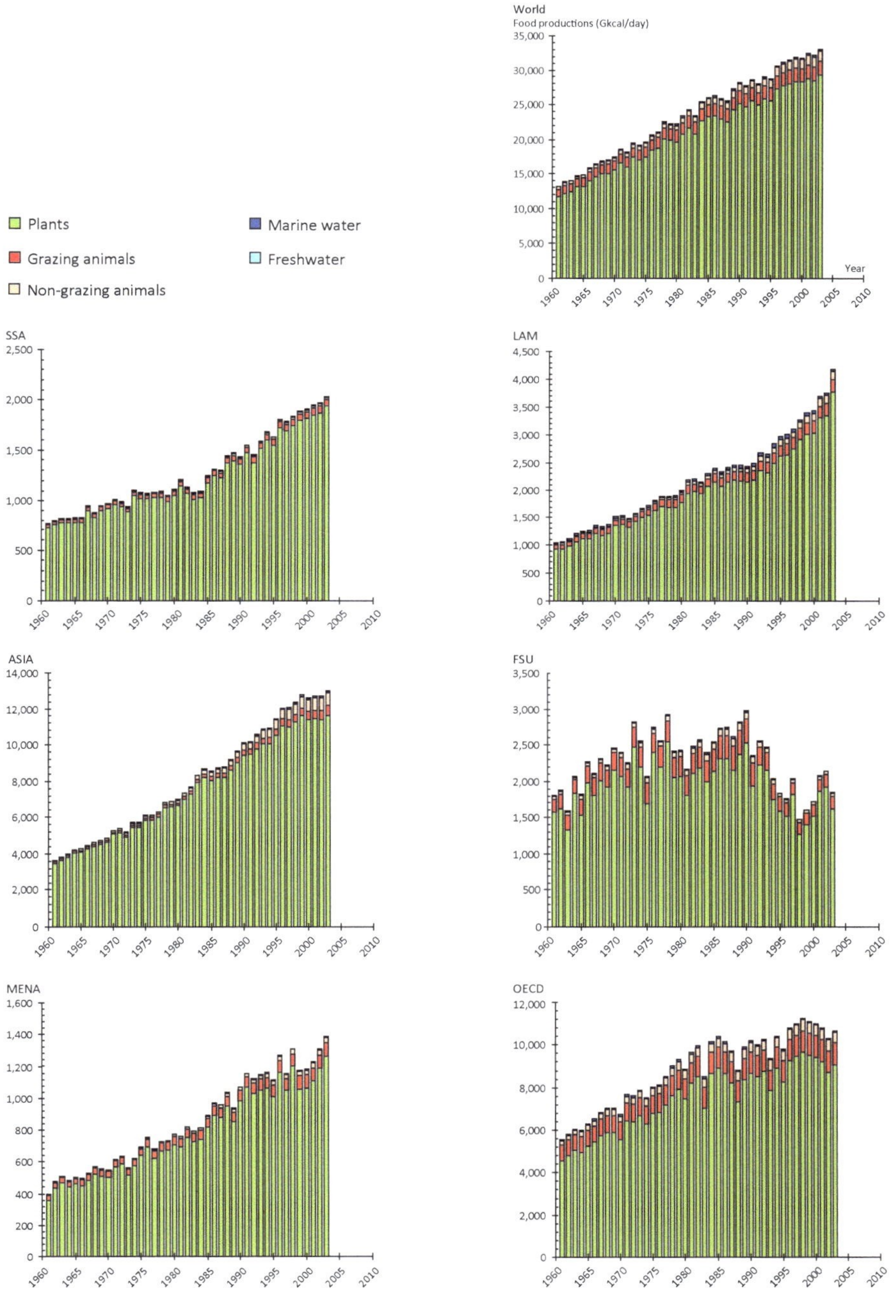

Fig. A2.5 Food productions (1961–2003). Gross production of food calories (Gkcal/day) under 5 categories of food origin: World estimates (Agribiom countries) and by MA region. (Source: B. Dorin, computed from FAO data)

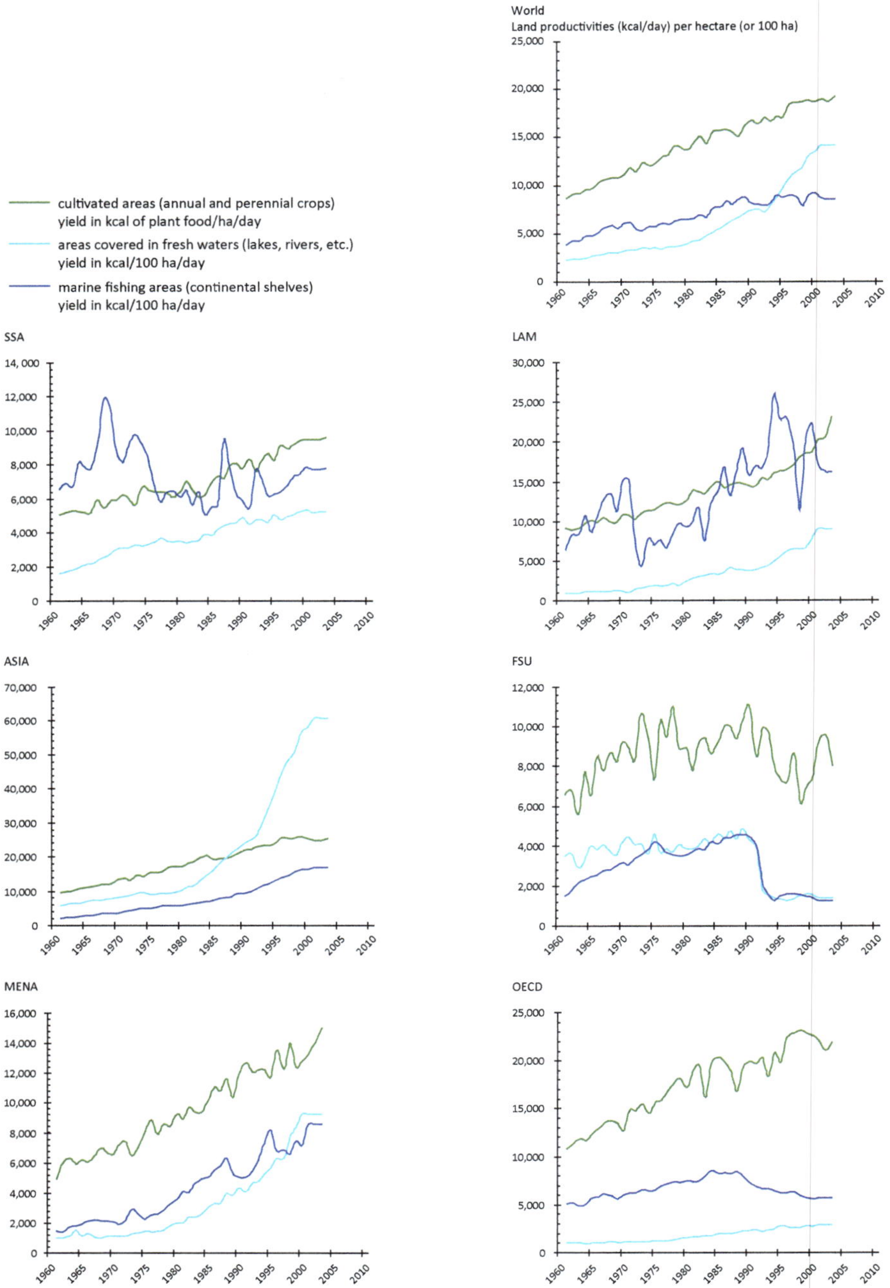

Fig. A2.6 Land productivities (1961–2003). Production of food calories (kcal/day) per hectare (or 100 ha) of 3 major areas: World estimates (Agribiom countries) and by MA region. (Source: B. Dorin, computed from FAO data)

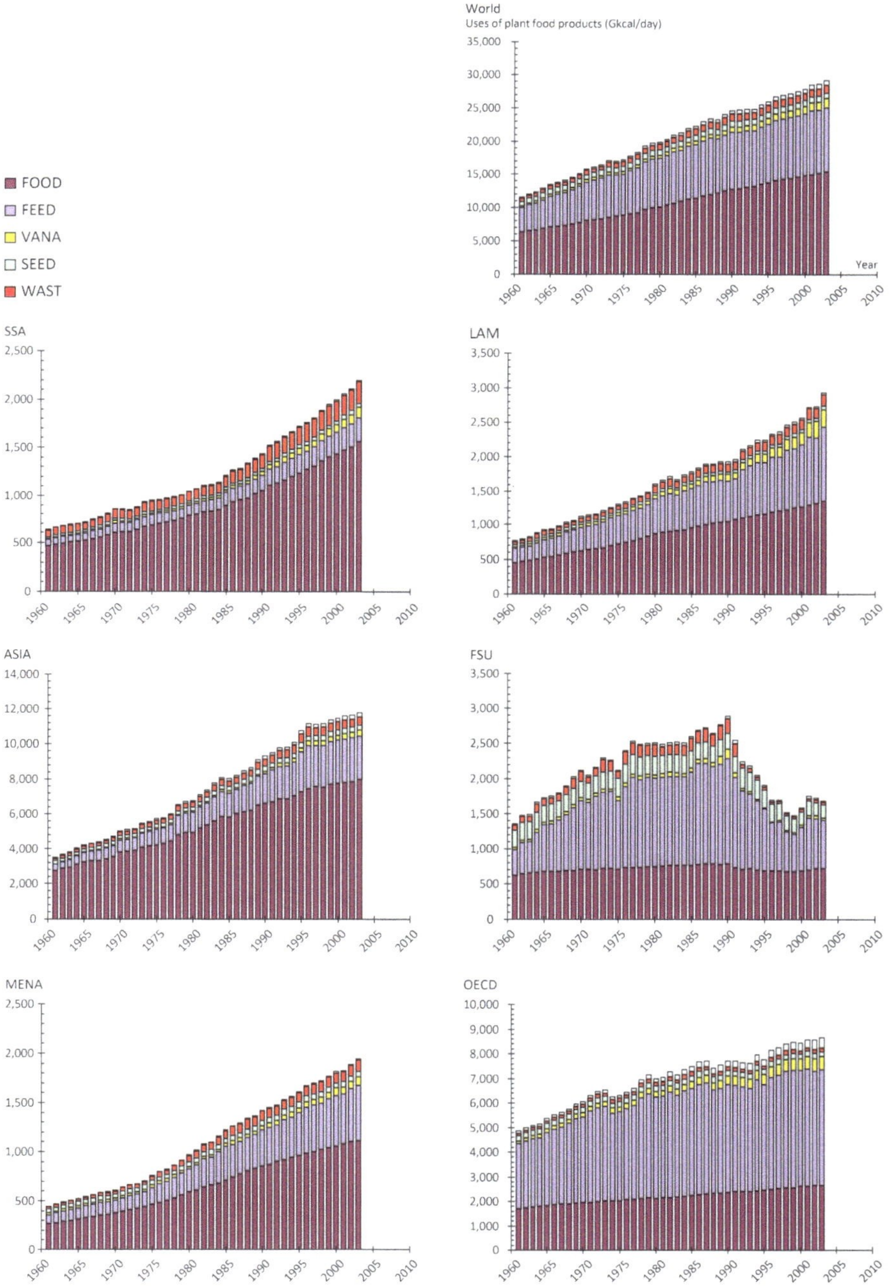

Fig. A2.7 Uses of plant food products (1961–2003). Use of plant food calories (Gkcal/day) under 5 categories of use: World estimates (Agribiom countries) and by MA region. (Source: B. Dorin, computed from FAO data)

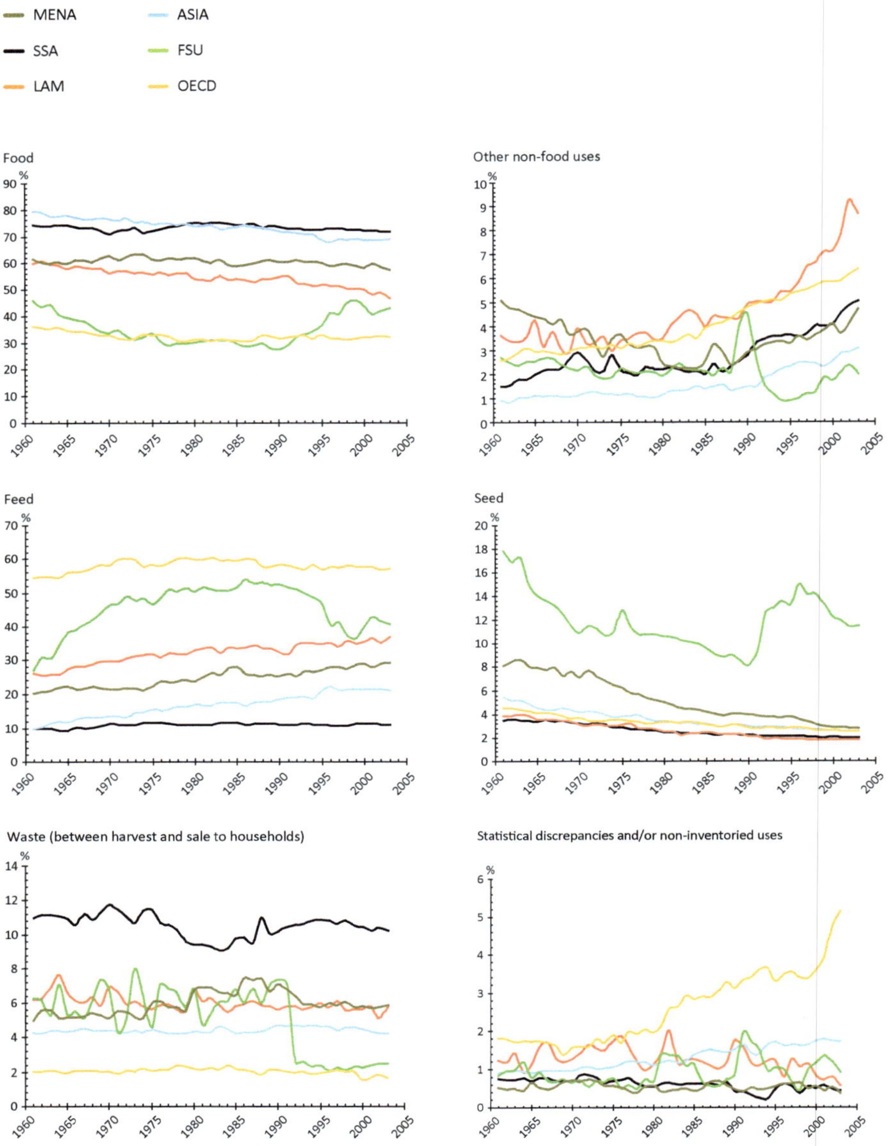

Fig. A2.8 Shares of plant food uses (1961–2003). Share (%) of various categories of plant food uses over total inventoried plant food use: *FOOD* human food, *FEED* animal feed, *VANA* other non-food uses, *SEED* seed, *WAST* waste (between harvest and sale to households); n.a., statistical discrepancies and/or non-inventoried uses Estimates by MA region (Agribiom countries). (Source: B. Dorin, computed from FAO data)

Fig. A2.9 Food Trade (1961–2003). Independence rate in food calories (%), i.e. Net trade of food (Exports—Imports, in Gkcal)/Total uses of food products (Gkcal), according to a few categories of food origin: World estimates (Agribiom countries) and by MA region. (Source: B. Dorin, computed from FAO data)

growth rates enabled us to calculate regional availabilities of aquatic calories for the Agrimonde GO scenario in 2050[1].

Plant Products

Plant calorie availabilities in the Agrimonde GO scenario correspond to the remaining calories needed to attain the level of total regional availabilities in the MA Global Orchestration scenario. They are obtained as follows:

$$\text{Plant availabilities} = \text{Total availabilities} - \text{Land animal availabilities} - \text{Aquatic availabilities}$$

Loss and Waste of Calories at the Different Stages Between Production and Final Consumption

Certain sources estimate vast amounts of loss: a global average of 30 % is estimated by (Smil 2000); this loss is distributed evenly between loss at the time of harvesting and loss at the retail and consumption stages.

This loss differs considerably between developed and developing countries:

- in the former, most waste is by consumers and the catering industry: up to 30 % for example in the US (Kantor et al. 1997) and the UK (WRAP 2008),
- in developing countries, most loss is in the fields (20–40 %) and then during transport and storage (Kader 2005) (Fig. A3.1).

Appendix 4

Land areas in Agrimonde GO

Land use statistics mobilised by the MA and Agribiom show three noteworthy differences:

- the land uses defined by the FAO, on which Agribiom draws, and those defined by the MA differ,
- the total surface area of emerged land per region also differs (a difference of 1–2 % exists between MA and FAO data),
- certain countries for which data are lacking or uncertain have been excluded in Agribiom.

---

[1] Note that the level of consumption of aquatic products is low compared to that of plant or land animal products. It never exceeds 2 % of the total calories consumed.

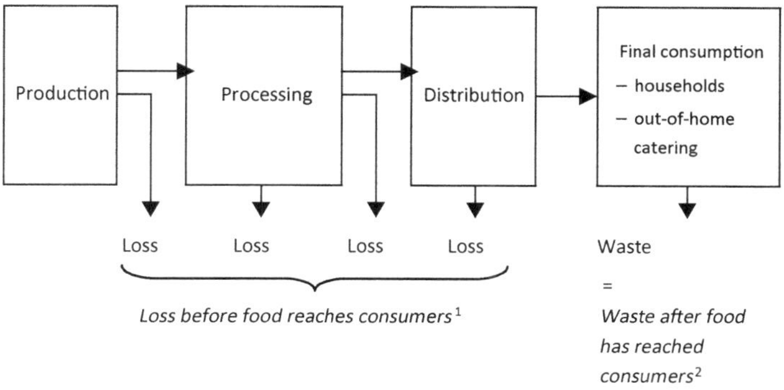

Fig. A3.1 Loss and waste of calories at the different stages between production and final consumption. *1* Loss *before reaching consumers* is recorded in FAO statistics. These figures include neither loss in the field, nor waste in the final consumption stage. The global average recorded by FAO in 2003 is 4 % of all use. This loss is explicitly taken into account in the assumptions made for calculating resource-use balances for 2050. As it is not analysed in depth in Agrimonde, it is represented in a highly conventional way. In Agrimonde 1 it never exceeds 4 %—In Agrimonde 1, it is assumed that the proportion of loss, before food reaches consumers, in total use in a given region was 4 % when its value in 2003 exceeded 4 %, and was maintained when its value in 2003 was lower than 4 % of total regional uses -, in Agrimonde GO the regional percentage of 2003 is maintained for 2050. *2* Waste *after reaching consumers* is included in food availability and there are no FAO statistics for this form of loss. In Agrimonde 1 the assumption of the reduction of this waste could be an important explanatory factor in managing the level of food availability per capita in 2050

Table A4.1 Land use in 2000 in Agribiom and in the MA (million ha)

Region	Cultivated areas[a] (Mha)		Pastures (Mha)		Forests (Mha)	
	MA	Agribiom	MA	Agribiom	MA	Agribiom
SSA	188	192	917	565	501	637
LAM	172	162	604	781	939	937
ASIA	456	454	645	359	459	497
FSU	216	203	320	327	1,271	843
OECD	417	418	732	752	1,295	979
World	1,543	1,512	3,505	3,339	4,483	3,926

[a] Including *NFCA* non-food cultivated area, *MA* millennium Ecosystem Assessment

The combination of these differences result in noteworthy discrepancies between total surface areas per region. In 2000, significant differences existed in cultivated areas between MENA, FSU and, to a lesser extent, LAM and ASIA. On average, the cultivated area in Agribiom is slightly smaller than that in the MA (Table A4.1). Total areas of pastures and especially of forests are much smaller in Agribiom.

To build the Agrimonde GO scenario based on the land use assumptions of the MA Global Orchestration scenario but on bases comparable to those of Agrimonde 1, the following corrections were made to the MA land surface areas:

Table A4.2 Land use in 2050 in the Agrimonde GO and MA Global Orchestration scenarios (million ha)

Regions	Cultivated areas (Mha)		Pastures (Mha)		Forests (Mha)	
	MA GO	Agrimonde GO	MA GO	Agrimonde GO	MA GO	Agrimonde GO
MENA	96	93	318	320	9	22
SSA	301	303	1,205	1,161	276	437
LAM	265	266	545	548	932	931
ASIA	498	504	726	735	386	442
FSU	219	223	208	212	1,389	945
OECD	457	467	595	608	1,428	1,110
World	1,836	1,856	3,596	3,584	4,421	3,887

- a regional corrective coefficient was applied, so that the total surface areas of the regions in the MA would correspond to those of Agribiom,
- the surface area of countries not taken into account in Agribiom was removed from the corresponding regions in the MA, taking into account current land use in these countries,
- the cultivated areas and pastures in 2050 in a particular region in the Agrimonde GO scenario were obtained by applying respectively the percentage of cultivated areas and of pastures to the total surface area of the region in 2003, in the MA Global Orchestration scenario,
- due to the high level of disparity between the MA and Agribiom definitions, the surface areas of forests are obtained by applying the above method as well as a regional corrective coefficient which most adequately reflects the evolution of forest areas between 2000 and 2050.

Table A4.2 below integrates these four levels of correction.

Appendix 5

Main Quantitative Assumptions of Agrimonde Scenarios Process for Attaining a Resource-Use Balance

Main Quantitative Assumptions of Agrimonde Scenarios

See tables on the following pages (Table A5.1 and A5.2).

Table A5.1 Assumptions related to the production of food calories

Variables		Regions	1961	2003	2050			
					Agrimonde 1		Agrimonde GO	
					Variant 1	Variant 2	Variant 1	Variant 2
Area (1,000 ha)	Cultivated area	OECD	426,495	415,865	495,000	495,000	467,000	467,000
		SSA	143,921	202,262	339,000	339, 000	303,000	303,000
		FSU	239,800	201,736	310,000	310,000	223,000	223,000
		ASIA	368,545	461,249	560,000	560,000	504,000	504,000
		LAM	102,362	163,882	310,000	310,000	266,000	266,000
		MENA	73,112	84,049	90,000	90,000	93,000	93,000
		Total	1,354,235	1,529,043	2,104,000	2,104,000	1,856,000	1,856,000
	Of wich non-food	OECD	0	0	95,000	95,000	66,000	66,000
		SSA	0	0	39,000	39,000	40,000	40,000
		FSU	0	0	10,000	10,000	36,000	36,000
		ASIA	0	0	20,000	20,000	28,000	28,000
		LAM	0	0	60,000	60,000	47,000	47,000
		MENA	0	0	200	200	56	56
		Total	0	0	224,200	224,200	217,056	217,056
	Pastures	OECD	816,819	736,014	576,000	576,000	608,000	608,000
		SSA	767,346	783,878	691,000	691,000	1,161,000	1,161,000
		FSU	302, 000	360,193	300,000	300,000	212,000	212,000
		ASIA	415,642	564,777	512,000	512,000	735,000	735,000
		LAM	461,731	553, 323	444, 625	444,625	548,000	548,000
		MENA	234,623	327,803	321,000	321,000	320,000	320,000
		Total	2,998,161	3,325,988	2,844,625	2,844,625	3,584,000	3,584,000
Area (1,000 ha)	Forest	OECD	1,070,679	980,752	1,077,644	1,077,644	1,109,987	1,109,987
		SSA	706,733	625,368	580,425	580,425	436,869	436,869
		FSU	913,000	842,734	843,000	843,000	945,234	945,234
		ASIA	526,033	499,932	445,434	445,434	441,770	441,770
		LAM	1,030,215	922,491	900,000	900,000	931 491	931,491
		MENA	49,193	33,501	32,965	32,965	21,616	21,616
		Total	4,295,853	3,904,776	3,879,469	3,879,469	3,886,966	3,886,966

Table A5.1 (continued)

Variables		Regions	1961	2003	2050			
					Agrimonde 1		Agrimonde GO	
					Variant 1	Variant 2	Variant 1	Variant 2
					22,600	22,600	33,507	33,507
		SSA	5,027	9,582	11,750	11,750	23,133	23,133
		FSU	6,549	8,026	14,500	14,500	12,825	12,825
		ASIA	9,485	25,251	25,100	25,100	46,416	46,416
		LAM	9,041	22,979	23,500	23,500	36,494	36,494
		MENA	4,921	15,010	14,500	14,500	21,362	21,362
		Total	8,607	19,189	20,027	20,027	32,942	32,942
Animal productions (Gkcal/day)	RUMI	OECD	711	1,068	519	1 034	1,401	1,735
		SSA	31	70	569	22	361	157
		FSU	185	160	88	551	291	323
		ASIA	89	500	773	232	1,635	1,481
		LAM	80	240	262	447	484	734
		MENA	33	93	230	26	230	27
		Total	1,128	2,132	2,441	2,312	4,402	4,457
	MONO	OECD	215	497	113	491	757	1,003
		SSA	4	23	283	0	167	61
		FSU	43	36	18	139	72	80
		ASIA	39	687	1,145	249	2,554	2,299
		LAM	19	137	169	331	350	568
		MENA	3	33	105	0	105	0
		Total	323	1,412	1,833	1,207	4,005	4,009

Table A5.2 Assumptions related to the use of food calories

Variables		Regions	1961	2003	2050			
					Agrimonde 1		Agrimonde GO	
					Variant 1	Variant 2	Variant 1	Variant 2
Population (1,000 inhab.)		OECD	743,048	986,872	1,066,211	1,066,211	1,066,211	1,066,211
		SSA	226,577	705,887	1,662,000	1,662,000	1,662,000	1,662,000
		FSU	217,854	279,012	239,212	239,212	239,212	239,212
		ASIA	1,510,658	3,322,361	4,427,101	4,427,101	4,427,101	4,427,101
		LAM	219,691	537,949	773,659	773,659	773,659	773,659
		MENA	128,242	371,745	631,964	631,964	631,964	631,964
		Total	3,046,070	6,203,826	8,800,147	8,800,147	8,800,147	8,800,147
Diet (kcal/hab./day)	VEGE	OECD	2,325	2,721	2,500	2,500	2,385	2,385
		SSA	2,057	2,218	2,500	2,500	2,667	2,667
		FSU	2,854	2,586	2,500	2,500	2,091	2,091
		ASIA	1,821	2,404	2,500	2,500	2,766	2,766
		LAM	2,069	2,528	2,500	2,500	2,758	2,758
		MENA	2,115	2,995	2,500	2,500	2,987	2,987
		Total	2,066	2,488	2,500	2,500	2,698	2,698
	RUMI	OECD	659	675	251	251	925	925
		SSA	121	102	129	129	214	214
		FSU	503	451	251	251	947	947
		ASIA	54	149	159	159	428	428
		LAM	297	357	251	251	507	507
		MENA	235	249	222	222	319	319
		Total	264	265	180	180	461	461
	MONO	OECD	282	512	200	200	703	703
		SSA	17	33	350	350	69	69
		FSU	196	213	212	212	349	349
		ASIA	25	209	253	253	443	443
		LAM	87	240	207	207	385	385
		MENA	21	96	238	238	123	123
		Total	104	233	259	259	373	373

Table A5.2 (continued)

Variables		Regions	1961	2003	2050			
					Agrimonde 1		Agrimonde GO	
					Variant 1	Variant 2	Variant 1	Variant 2
Total food uses (Gkcal/day)	VEGE	OECD	5,164	8,669	4,856	8,590	10,939	13,357
		SSA	632	2,191	7,515	4,426	7,378	6,084
		FSU	1,251	1,580	1,017	3,900	2,118	2,318
		ASIA	3,695	11,807	16,732	13,554	23,009	22,094
		LAM	954	3,109	3,977	5,425	5,930	79,13
		MENA	493	1,985	3,549	1,752	4,176	2,223
		Total	12,190	29,341	37,646	37,646	53,551	53,990
	RUMI	OECD	730	953	445	445	1,425	1,425
		SSA	34	82	243	243	405	405
		FSU	171	171	66	66	290	290
		ASIA	61	542	795	795	2,208	2,212
		LAM	97	267	273	273	534	534
		MENA	43	117	178	178	254	254
		Total	1,135	2,132	2,000	2,000	5,116	5,120
	MONO	OECD	183	484	187	187	733	729
		SSA	7	26	609	608	122	122
		FSU	33	49	40	40	73	73
		ASIA	19	688	1,122	1,122	1,980	1,980
		LAM	13	127	158	159	301	302
		MENA	2	37	157	157	81	81
		Total	257	1,412	2,274	2,273	3,291	3,287
Of which feed (Gkcal/day)	VEGE	OECD	2,608	4,712	1,311	4,672	6,911	9,093
		SSA	62	245	2906	0	1,720	639
		FSU	365	686	350	2,761	1,370	1,537
		ASIA	342	2,449	3,896	1,033	8,324	7,509
		LAM	199	1,078	1,328	2,586	2,730	4,427
		MENA	90	561	1,706	0	1,700	0
		Total	3,665	9,731	11,497	11,052	22,755	23,206

Table A5.2 (continued)

Variables		Regions	1961	2003	2050			
					Agrimonde 1		Agrimonde GO	
					Variant 1	Variant 2	Variant 1	Variant 2
Of which non-food use (Gkcal/ day)	VEGE	OECD	123	528	284	497	635	775
		SSA	9	110	0	0	301	249
		FSU	36	34	17	68	36	39
		ASIA	31	353	419	335	574	545
		LAM	27	252	283	385	421	561
		MENA	22	91	0	0	171	91
		Total	249	1,368	1,003	1,284	2,138	2,260

Process for Attaining a Resource-Use Balance in the Two Variants

Stages in Attaining a Balance in Variant 1

Stage 1: Calculation of Needs Plant calorie requirements for animal feed are calculated so that a region's production of animal calories is equal to its needs. Production functions are used to deduce the quantity of feed needed to achieve the targeted animal production (exogenous variable) given the area under pastures (exogenous variable).

All the other needs in plant and animal calories were set as exogenous variables by the panel.

Stage 2: Achieving Regional Balances Through Trade Plant calories

There are two possibilities:

Animal calories

The calorie balance is attained in Stage 1. Each region produces the exact amount of calories that it uses.

- the region's production of plant calories does not cover all of its needs; it therefore imports plant calories corresponding to the difference between all of its needs (defined above) and its initial plant resources,
- the region's production covers its needs; it can export calories corresponding to the difference between its plant production and its regional needs as defined above.

Stage 3: Global Adjustment The global trade balance serves to check whether total use corresponds to total resources. If the needs and resources of each region are balanced—after trade—then there will be a global balance and the quantitative assumptions of the scenario will be coherent. The quantity of exportable calories is equal to the need for imports. Otherwise, the global situation may be discussed by the panel (the scenario has a surplus or deficit).

Stages in attaining a balance in Variant 2

The stages in attaining the balance are as follows.

Stage 1: Calculating Calorie Needs Needs in plant and animal calories are all exogenous variables, except for plant calorie needs for animal feed.

Stage 2: achieving regional balances through trade There are three possibilities:

- the region's plant production does not cover its plant calorie needs for human food. In this case, the region imports the plant calories corresponding to its human needs (all uses except animal feed). Its animal production is calculated by the production functions using pastures (exogenous variable in the scenario) but not other plant calories. It imports the animal calories that it lacks to meet its needs,

- the region's plant production covers its human plant calorie needs but not its needs in animal feed (to meet the regional population's animal calorie needs). The region therefore imports animal calories to cover these needs,
- the region's plant production is enough to cover its human needs in plant calories and its needs in animal feed (to cover the population's needs in animal calories).

The region therefore exports (together with all exporting regions, and in the same quantity) the plant calories needed to satisfy the needs of those regions that import plant calories, and uses all the rest for its animal production. Excess animal production is exportable.

Stage 3: Global Balance The global trade balance enables us to check whether the total of all use corresponds to the total of all resources. If each region's needs and resources—after trade—are balanced, then there is a global balance, and the scenario's quantitative assumptions are coherent. Otherwise, the global situation may be discussed by the panel (the scenario has a surplus or a deficit).

Appendix 6

Assumptions on the Qualitative Dimensions of the Agrimonde Scenarios

Within the framework of the dimensions and variables of the Agrimonde system presented in Chap. 1 (Table 1.1), the first dimension groups together variables of a contextual and global nature, while the other dimensions include variables likely to have a more direct impact since they are situated closer to the heart of agricultural and food systems. This framework enabled us to explore in greater detail the qualitative dimensions left open in the first steps of scenario-building [scenario-building principles (Chap. 4), quantification (Chaps. 5 to 8), and analysis in terms of comparison, coherence and drivers of change (Chap. 9)]. For Agrimonde GO, apart from the assumptions associated with demographic trends (which are the same as in the Agrimonde 1 scenario, causing demographic pressures to be similar), the assumptions on the qualitative variables are based on our understanding of the Global Orchestration scenario as proposed in the MA report (MA 2005b).

The Global Context

Strong Growth in Both Scenarios, but Different Urban and Rural Population Dynamics

In the scenario-building carried out in the preceding chapters, identical assumptions are made on the world's population in the two Agrimonde scenarios (some 9 billion inhabitants in the world in 2050). The assumptions on urbanisation and the rural

Table A6.1 Global context

	Agrimonde GO	Agrimonde 1
Population	9 bn inhabitants (half in ASIA)	9 bn inhabitants (half in ASIA)
Urbanisation and rural exodus	Acceleration	Stabilisation
Economic growth	Very strong	Strong, driven by growth in developing countries
Advances in knowledge	Very fast	Very fast
Income distribution	More equitable owing to the decline in rural poverty	Far more equitable owing to decline in rural poverty and reduction of North-South inequalities
Agricultural commodity prices	Wheat and maize: increase	Real prices on an upward curve
	Rice and animal products: decrease[a]	With high level of volatility at the beginning of the period

[a] These evolutions are a result of MA simulations

exodus, economic growth and income distribution on a global scale are limited to some extent by the scenario-building principles and the quantitative assumptions on agricultural resources and uses. In Agrimonde 1, the pace of urbanisation has reached a stabilisation phase. In this sense it differs from Agrimonde GO, in which it is accelerating as a result of very rapid technological progress in agriculture, geared towards capital-labour substitution.

In Agrimonde 1, agricultural development, coupled with increased food consumption in kilocalories in the regions currently situated below the world average, seems complementary to strong global economic growth and more equitable income distribution. In the OECD region, whose mean consumption in kilocalories declines by a quarter, an assumption of 'degrowth' would have been feasible. However, given the pace of the other regions' development and economic take-off, it seems more coherent to imagine that the extent of markets for industrial, service and agricultural products (as the regions of ASIA, MENA, and SSA still have shortages) drives the growth of the rich countries. Moreover, the drop in the mean level of kilocalories consumed in this region is associated with an improvement in the quality of the products consumed. It is therefore compatible with stable or even increasing average household food budgets. Very strong world economic growth is an assumption of the Agrimonde GO scenario. It stems from rapid technological progress and trade liberalisation, which are assumed to allow for economic take-off and subsequently a fairer distribution of the world's income (Table A6.1).

The pace of progress in knowledge, very rapid in Agrimonde GO, seems to be equalled in Agrimonde 1. This has resulted in an improvement in market and institutional infrastructures, higher levels of qualifications, and innovation diffusion in developing countries. It has also allowed for improvements in energy efficiency and the development of renewable energies as substitutes for fossil fuels, to limit greenhouse gas emissions. More generally, as the protection of the environment and natural resources is a high priority in this scenario, a multitude of innovations must be developed to allow for more sustainable modes of production of goods and services.

The mode of quantification of the Agrimonde scenarios does not allow for agricultural price trends to be determined, as resource-use balances are not market equilibriums. Demographic pressure and the increase in the mean per capita income implied by the quantitative scenario, on the demand side, and the type of technological progress foreseen (innovations to meet multiple objectives, rather than the central objective of increasing yields), on the supply side, point to increasing agricultural prices in the medium-long term. Moreover, it seems likely that the repetition of food crises such as that of 2008 would be one of the factors of emergence of the Agrimonde 1 scenario. This scenario could therefore be characterised by strong price volatility at the beginning of the period. For Agrimonde GO, the work of the MA experts, which was based on economic modelling, is intended to be more precise as regards price trends. While the tensions related to demand result in increases in the price of wheat and maize, technological progress allows for a decrease in the price of rice and animal products.

International Regulations

Strong International Cooperation To Favour Trade, in Both Scenarios, Coupled with Ambitious Environmental Regulations in Agrimonde 1

Even if they are relatively open in the Agrimonde 1 scenario, international political relations have to allow not only for extensive trade in agricultural products (more than in Agrimonde GO) but also for highly ambitious environmental regulations (weak in Agrimonde GO). Thus, irrespective of the scenario, they are characterised by a high level of international cooperation. This cooperation may be driven by a dominant actor or else result from a multipolar geopolitical configuration. In both scenarios international regulations have to be accompanied by massive North-South capital flows, as development is one of the priorities.

In Agrimonde 1, the regulations of agricultural trade must:

- prevent price distortions unfavourable to the development of agriculture in developing countries,
- allow for temporary exemptions for countries whose development is based essentially on agriculture,
- make it possible to reveal the environmental costs associated with agricultural activities, to encourage farmers to develop more sustainable farming systems.

The scenarios do not specify the modalities of implementation of international regulations. In Agrimonde 1 we can foresee an organisation of international trade similar to those of the early twenty-first century. Alternatively, it may be radically different with, for example, the creation of an agricultural trade organisation responsible for guaranteeing food security.

Table A6.2 International Regulations

	Agrimonde GO	Agrimonde 1
International political relations	Multilateral cooperation prevails	Strong international cooperation
Organisation of international trade	Liberalisation	Liberalisation but significant exceptions both for agricultural countries and to preserve the environment
International agreements on climate	None	Ambitious
International agreements on biodiversity	None	Ambitious
Governance and management of sanitary risks	Effective owing to global coordination and technological progress	Effective owing to global coordination and the resilience of ecosystems
Governance and management of marine resources	In reaction to ecological crises	Proactive and effective
North-South capital flows	Significant	Significant

The Dynamics of Agricultural Production

Production, at the heart of the Agrimonde system, is strongly limited by the scenario-building principles and quantitative assumptions. On the other hand, in both scenarios the dynamics of agri-food industries are left open (Table A6.2).

Scenarios of Highly Dynamic Agricultures, Distinguished by Land Use Patterns and Social Forms of Production

Total areas under agricultural production (cultivated land and pastures) have increased more in Agrimonde GO than in Agrimonde 1. Although in Agrimonde 1 the cultivated areas have increased more, this is largely offset by the reduction of pastures. As noted in Chap. 9, Agrimonde GO and Agrimonde 1 represent two very different strategies to implement the trade-off between increasing cultivated areas and improving yields. In Agrimonde GO, cultivated areas increase more moderately than in Agrimonde 1 but yields increase faster owing to production technologies that make it possible to substitute capital for labour and to substantially increase production per hectare. In Agrimonde 1 production technologies are based on ecological intensification; they make it possible to maintain or even to increase yields while strongly limiting dependence on fossil fuels, the use of inputs, and consequently the impacts of agricultural activities on ecosystems.

In both scenarios, investments in agricultural production at the farm level as well as the infrastructure level have increased steeply, especially in developing countries. Whereas in Agrimonde GO fairly large investments have been made in irrigation, in Agrimonde 1 this applies only in SSA, for production technologies in this scenario favour water conservation in ecosystems and suitable cropping patterns. In

Table A6.3 Dynamics of agricultural production

	Agrimonde GO	Agrimonde 1
Production areas	Maintenance of pastures and extension of cultivated areas (two-thirds biofuels)	Steep reduction of pastures and significant extension of cultivated surfaces (just over one third with biofuels)
Investments in farming	Heavy investments, especially in irrigation	Heavy investments in developing countries
Investments in infrastructures and public goods	Heavy investments, especially in developing countries and for irrigation	Heavy investments in developing countries
		Geared towards multifunctionality in rich countries
Social forms of production	Strong presence of capitalistic forms	Diversity
		Strong component of peasant and family farming in developing countries
Production techniques	Intensification, technological standardisation, and strong development of GMOs	Ecological engineering, biotechnologies, local adaptation
Processing (agro-industry): organisation and production technologies	Concentration, automation of processes, strive for economies of scale	Diversity of entrepreneurial forms—Search for economy of variety, valorisation of co-products and waste

this scenario the multifunctionality of agriculture in rich countries is accompanied by major investments in landscape management, the prevention of natural risks and, more generally, the collective management of natural resources (Table A6.3).

The scenario-building principles imply highly contrasting social forms of production in the two scenarios. While Agrimonde GO is characterised by the industrialisation of agriculture, Agrimonde 1 is based on more varied social forms of production.

The Dynamics of Agri-Food Industries are Decisive for the Future of Food and Agriculture but are Difficult to Apprehend through Quantitative Scenarios Alone

The two scenarios remain vague as regards industrial organisation and production technologies in the food-processing sector, even though these aspects will be decisive for the future of agricultural and food systems. While concentration, process automation, and the strive for economies of scale appear to be consistent with the spirit of the Agrimonde GO scenario, Agrimonde 1 is more a scenario of diversity of entrepreneurial forms, where SMEs, cooperatives and multinational firms coexist. Production technologies are oriented more towards economies of variety or a valorisation of co-products and waste.

Dynamics of Biomass Consumption

Dietary Trends: Continuation in Agrimonde GO, Radical Changes in Agrimonde 1

Agrimonde GO is intended to be a trend-based scenario as regards diet, since food consumption, especially food of animal origin, increases with income. Agrimonde 1, on the other hand, foresees major changes in diet linked to environmental and above all nutritional concerns, with the struggle against obesity being a key objective. While the quantitative assumptions concern only the mean number of kilocalories consumed in the different regions, and their distribution in terms of origin (plant, non-grazing animals, grazing animals, aquatic), it seems probable that the strong shifts marking them are also accompanied by major changes in food consumption practices and, more generally, lifestyle, especially in those regions that experience a decrease in total calorie consumption. We can consider that Agrimonde 1 mainly represents a scenario in which consumers in rich countries reinvest time in the preparation of meals, buy more raw products in shorter distribution channels, and so on. However we may also see the catering industry as an ideal means for changing food-related behaviours and disseminating messages on nutritional policies.

In the world of the Agrimonde GO scenario, citizens trust in science to control sanitary and environmental risks, whereas in the scenario-building principles of Agrimonde 1 no reference is made to citizens' awareness of sanitary issues. Citizens in Agrimonde 1 are however keenly aware of environmental protection as it is a priority. This is reflected in their consumption behaviours and in the pressure they bring to bear on public policy-makers.

Energy and Food Needs do not Compete in Either of the Scenarios, but for Different Reasons

In both scenarios the demand for biomass for energy purposes has increased substantially by 2050 compared to the beginning of the century[2]. It has not however entered into competition with the food demand, even though the reasons differ in the two scenarios. In Agrimonde GO, yield increases have made it possible to satisfy both types of need. In Agrimonde 1 only regions with an agricultural potential allowing for positive resource-use balances produce biofuels for the energy market. Not only do the other regions produce few biofuel crops, but when they do it is for the farms' own energy autonomy. Highly contrasting assumptions can be made concerning the industrial consumption of biomass (excluding energy) in the two scenarios, even though Agrimonde 1 corresponds essentially to a scenario in which the search for substitutes to oil encourages the replacement of traditional carbochemistry by biomass-based carbochemistry (Table A6.4).

[2] At global level areas used for biofuel crops are similar in both Agrimonde scenarios. Compared to the other MA scenarios, the biofuel areas in Global Orchestration are greater than those of the Order from Strength and TechnoGarden scenarios, but less than those of Adapting Mosaic.

Table A6.4 Dynamics of biomass consumption

	Agrimonde GO	Agrimonde 1
Consumption habits and diets	Steep increase in total calorie intake and in the consumption of meat and fish	Major changes, especially related to poverty alleviation and, for the more wealthy, nutritional concerns
Society's awareness of sanitary issues	Trust in science's ability to find solutions	
Society's awareness of environmental issues	Trust in science's ability to find solutions	Acute awareness: the environment is a social priority
Consumption of biomass for energy production	High	High, geared towards the autonomy of farms in regions with a shortage of food calories
Consumption of biomass for the production of industrial goods		Biomass-based carbochemistry gradually replaces petrochemistry

The Actors' Strategies

Trade Encouraged Irrespective of the Scenario and Notwithstanding Contrasting Political Philosophies

The public and private actors' strategies in the field of agriculture and food are determining factors in both scenarios. In Agrimonde GO, support for agricultural production has gradually waned, in compliance with international trade agreements. Public action is generally reactive, whether it concerns nutrition, the environment, or energy (its only reason to encourage energy efficiency gains is to cope with the scarcity of fossil fuels).

In Agrimonde 1, trade in agricultural products must be strongly encouraged. Direct support for production is therefore destined to disappear (it was allowed only during the agricultural take-off phase in those countries most dependent on agriculture). The liberalisation of trade has not however been accompanied by less government intervention. Public intervention has been decisive and proactive, and aimed at regional development, protection of ecosystems, and climate change adaptation and mitigation. In this respect, an assumption of ambitious policies promoting renewable energies on a decentralised basis would be fairly coherent in Agrimonde 1, especially in countries where energy access problems are currently an obstacle to development, but also in cities, through waste recovery or energy-autonomous buildings. As regards transport, we can imagine that by 2050 the use of electric vehicles will be standard practice owing to massive investments in public research and to policies to support private investments, with the aim of removing the technical and economic obstacles related to fuel cells (cell durability, sustainable production, safe distribution and hydrogen storage, etc.). Finally, nutrition-related policies are also highly ambitious in Agrimonde 1, but their modalities are yet to be explored.

They have been highly innovative compared to the beginning of the century, and have impacted significantly on diet and more generally on lifestyles.

More Balanced Influence of Private Actors in Agrimonde 1

In Agrimonde GO large multinational corporations increase their weight in the agri-food value chain, while the other actors likely to weigh on public policies—such as professional agricultural organisations or NGOs (non-governmental organisations)—are more in the background, even though NGOs have been key players in making development a top priority on international policy agendas. The influential power of private actors seems to be more balanced in Agrimonde 1. This is because the agri-food sector has not experienced the concentration trend specific to Agrimonde GO, as a variety of actors co-exist (in terms of products offered, and of size and type of enterprise). Moreover, while professional agricultural organisations have seen their power grow substantially in developing countries (to allow for a rebalancing of the respective influence of city-dwellers and rural populations on policies), the power of agricultural organisations has probably been counteracted by environmental NGOs in rich countries.

Knowledge and Technologies in the Field of Agriculture and Food

Scenarios of Major Efforts in Research and Innovation, for DIfferent Purposes and with Specific Modes of Knowledge Production and Dissemination (Table A6.5)

In these two scenarios the public- and private-sector research and innovation effort in the food and agricultural field has had to be massive and largely international. It has been complementary to heavy investments in training farmers in developing countries. Global food security is a major challenge in both scenarios and relies on the valorisation of the diversity of the world's agricultural potential. While in Agrimonde GO the objective of innovations is primarily to increase yields, in Agrimonde 1 these increases must be compatible with the objectives of protecting ecosystems and reducing dependence on inputs. In Agrimonde 1, strong incentives are provided in this respect, through national and international framework policies for public and private research. Ecological intensification, with the decisive progress in knowledge on ecosystems that it implies, is based on a transformation of modes of production and dissemination of knowledge It necessitates extensive training for farmers in developing regions and rich countries alike (Table A6.6).

The Agrimonde 1 scenario remains vague as regards knowledge to be developed in the field of nutrition, even though, as we have seen, such knowledge is essential for its credibility. Yet it seems that in Agrimonde 1 the challenge represented by the struggle against obesity has been met owing to breakthroughs in knowledge, including on food-related behaviours, to back up public policies. Beyond the nutritional field as such, Agrimonde 1 assumes that organisational innovations in the

Table A6.5 Actors' strategies

		Agrimonde GO	Agrimonde 1
States' strategies	Agricultural policies	Substantial decrease in support	Substantial decrease in support for production but agrarian reforms and tariff protection for local produce in countries highly dependent on agriculture Remuneration of environmental services (multifunctionality and regional development)
	Sanitary and nutrition policies	Reactive, especially with regard to the obesity epidemic	Highly active and effective
	Energy policies	Search for better energy efficiency	Highly active: R&D, substitution of renewable energies for fossil fuels; energy efficiency
	Environmental policies	Reactive	Proactive: coupled with development and regional planning policies
Private actors' strategies	Role of professional agricultural organisations		Considerable, especially in developing countries
	Strategies of multinational firms	Multinational firms increase their control over agricultural production	High level of segmentation of markets and diversity of actors
	Role of NGOs	Important role of NGOs in development	Important role of development and environmental protection NGOs

processing-distribution channels have led to significant reductions in losses and waste at that level.

In Agrimonde GO the strengthening of intellectual property rights (IPR), including on living organisms, is one of the factors at the origin of the research and innovation effort. It is nevertheless challenged by the developing countries, as described in the Global Orchestration scenario proposed by the MA experts. In Agrimonde 1 the quantification of the assumptions and the scenario-building principles do not enable precise conclusions to be drawn on the evolution of IPR. The scenario nevertheless seems to be consistent with larger possibilities of exemptions to cope with major public health, environmental or food security problems, when the impossibility of acquiring licences threatens agricultural development capacities.

Table A6.6 Knowledge and technologies in the field of food and agriculture

	Agrimonde GO	Agrimonde 1
Investments in public and private R&D	Heavy investments (public and private)	Heavy investments (public and private), oriented by public policies
Objectives of innovations	Yield gains	Ecological intensification
Intellectual property system for living organisms	Strengthening IPR is starting to be called into question by developing countries at the end of the period	IPR systems with strong exemptions (public health, development, environment)
Orientations of agricultural research	Genetic engineering, agro-chemistry, irrigation techniques, etc.	Knowledge of the functioning of ecosystems (ecology, genomics), ecological engineering, biotechnologies
Farmers' training	Developed but with little focus on environmental management; standardisation of skills	Highly developed in developing countries and in rich countries
Organisation and actors of innovation and of its diffusion	Public and private research laboratories, centres of agricultural training	Multiple (researchers, trainers, professionals) and interactive (clusters, highly internationalised communities of practice and epistemic communities)

Sustainable Development

Reducing Poverty and Protecting the Environment: A Trade-Off in Agrimonde GO; Synergies and Compromises in Agrimonde 1

In both scenarios, decisive progress has been made in alleviating poverty and malnutrition. In Agrimonde GO this objective has been met to the detriment of other objectives such as the protection of ecosystems, the struggle against climate change, or bringing the obesity epidemic under control. In contrast, Agrimonde 1 explores the complementarity of these objectives. However, the protection of ecosystems in certain areas risks limiting agricultural development, as seen in SSA. Limited yield gains have been offset by a considerable expansion of cultivated areas, which raise sustainability problems as far as greenhouse gas emissions and biodiversity are concerned.

In Agrimonde GO the high economic growth has resulted in an explosion of the energy demand. This demand is satisfied above all by fossil fuels (including coal), even though the use of biofuels is increasing and renewable energies account for a total of 10 % of the energy consumed in 2050. Consequently, it is in this MA scenario that climate change is most marked. Unlike Agrimonde GO, the struggle against climate change is a priority in Agrimonde 1 and massive investments in the development of new energy sources have made it possible to limit greenhouse gas emissions. The fact remains that by 2050 the inertia associated with climate change, coupled with still fragmented knowledge on this phenomenon and its consequences,

Table A6.7 Sustainable development

		Agrimonde GO	Agrimonde 1
Natural resources	Biodiversity conservation	Deterioration	Loss of wild biodiversity; gain in domestic biodiversity
	Greenhouse gas emissions and climate	Very steep increase in greenhouse gas emissions (+50 %)	After a peak of emissions in 2020, decline of greenhouse gas emissions below the 2000 level
		Very steep increase in global temperatures (highest in the 4 MA scenarios)	Acceleration of climate change in the first quarter of the century; the effects of mitigation policies only felt towards 2050
	Soil fertility	Deterioration due to chemical products and agricultural practices	Decrease of erosion and salinisation owing to ecological intensification
	Water (availability and quality	Increased use of water (owing to more availability as a result of climate change)	Improvement of the water-related services of ecosystems
		Deterioration of the quality	Better management of the resource
			Weak development of irrigation
Social equity	Satisfaction of essential needs (food, health, employment, education)	Improvement with a reduction in inequalities	Improvement with a reduction in inequalities
	Quality of life: dwellings, culture, social relations	Improvement but problems of sustainability of megalopoles, pollution, etc.	Improvement with promotion of cultural diversity—New town-country relations

make it impossible to affirm that global warming will be worse in Agrimonde GO than in Agrimonde 1. As a rupture scenario that is highly demanding as regards public actions, Agrimonde 1 is the type of scenario which emerges in response to crises. Accelerated climate change in the early twenty-first century could therefore be one of the crises at the origin of this scenario. Even though other conjectures can be considered, we can also assume that the active struggle against climate change will have started to have effects on the climate in around 2050 (Table A6.7).

References

Altieri MA (1987) Agroecology: the scientific basis of alternative agriculture. Westview Press, Boulder, 227 p

Basdevant A (2003) Obésité humaine: origine et conséquences. In: Basdevant A, Ricquier D (eds) Pour une approche scientifique de l'obésité. Annales de l'Institut Pasteur-Actualités, Elsevier, Paris, pp 11–26

Bhadwal S, Anantram K, Nair S, Kumar V, Kulkar U (2006) Mapping climate vulnerability and poverty in Africa. Report to the Department for International Development, International Livestock Research Institute, Nairobi, Kenya, 171 p

Bouwman AF, Hoek KWVd, Eickhout B, Soenario I (2005) Exploring changes in world ruminant production systems. Agric Syst 84(2):121–153

Braun J von, Rosegrant MW, Pandya-Lorch R, Cohen MJ, Cline SA, Brown MA, Bos MS (2005) New risks and opportunities for food security. Scenario Analyses for 2015 and 2050. International Food Policy Research Institute, 2020 Discussion Paper, 39, 32 p

Brown L (1995) Who will feed China: wake-up call for a small planet. Worldwatch environmental alert series. W.W. Norton & Company, New York, 163 p

Caron P, Reig E, Roep D, Hediger W, Le Cotty T, Barthélémy D, Hadynska A, Hadynski J, Oostindie H, Sabourin E (2008) Multifunctionality: refocusing a spreading, loose and fashionable concept for looking at sustainability? Int J Agric Resour Gov Ecol 7(4–5):301–318

Challinor AJ, Wheeler TR, Garforth C, Craufurd P, Kassam A (2007) Assessing the vulnerability of food crop systems in Africa to climate change. Clim Chang 83:381–399

CIHEAM (2008). Mediterra 2008—the future of agriculture and food in mediterranean countries. International Centre for Advanced Mediterranean Agronomic Studies, Presses de Sciences Po, Paris, 360 p

CIRAD-GRET-MAE (2002) Mémento de l'agronome, Centre de coopération internationale en recherche agronomique pour le développement, Groupe de recherche et d'échanges technologiques, Ministère des affaires étrangères, Quae, 5e édition, Versailles, 1 692 p

Cline WR (2007) Global warming and agriculture: impacts estimates by country. Peterson Institute for International economics, Washington, DC, 250 p

Collomb P (1999) Une voie étroite pour la sécurité alimentaire d'ici à 2050, Economica, Paris, 197 p

Combris P, Amiot-Carlin M-J, Caillavet F, Causse M, Dallongeville J, Padilla M, Renard C, Soler LG (2008) Les fruits et légumes dans l'alimentation. Enjeux et déterminants de la consommation. Collection Expertises collectives, Éditions Quae, Versailles, 128 p

Commissariat général du Plan (2004) La méthode des scénarios, outils d'une démarche prospective. *Les dossiers d'Aleph* 1:16

Conway G (1998) The doubly green revolution: food for all in the twenty-first century. Penguin Books, London, 335 p

Cosgrove WJ, Rijsberman FR (2000) World water vision: making water everybody's business, world water council. Earthscan Ltd, London, 108 p

S. Paillard et al. (eds.), *Agrimonde – Scenarios and Challenges for Feeding the World in 2050*, DOI 10.1007/978-94-017-8745-1,

Crombie LK, Irvine L, Elliott L, Wallace H (2005) Public health policy to tackle obesity: an international perspective. University of Dundee, NHS Scotland, 73 p
D'Arcy M, Harduar P, Orloff M, Rozas AK (2006) The move towards obesity: the nutrition transition in China, Paper prepared for the International Economic Development Program, Ford School of Public Policy, University of Michigan, 28 p
Deaton A, Dreze J (2009) Food and nutrition in India: facts and interpretations. Econ Polit Wkly 44(7):42–65
De Jouvenel H (2000) A brief methodological guide to scenario building. Technol Forecas Soc Chang 65(1):37–48
Dekhili S (2004) Contribution des politiques de sécurité alimentaire au bien-être des populations: le cas de la Tunisie. Mémoire de recherche de DEA EGDAAR, Université Montpellier I, SupAgro-Montpellier
Delgado C, Rosegrant M, Steinfeld H, Ehui S, Courbois C (1999) Livestock to 2020. The next food revolution. Food, Agriculture, and the Environment Discussion Paper 28, International Food Policy Research Institute, Food and Agriculture Organization of the United Nations, International Livestock Research Institute, Whashington, DC, USA, 72 p
Delpeuch F, Maire B, Monnier E, Holdsworth M (2009) *Globesity: a planet out of control*? Earthscan Ltd., London, 178 p
Dériot G (2005) La prévention et la prise en charge de l'obésité, Rapport de l'Office parlementaire d'évaluation des politiques de santé, 8 (2005–2006), 204 p
Devendra C, Sevilla CC (2002) Availability and use of feed resources in crop-animal systems in Asia. Agric Syst 71(1–2):59–73
Dorin B (1999) Food policy and nutritional security. The unequal access to lipids in India. Econ Polit Wkly 34(26):1709–1717
Dorin B, Gitz V (2008) Écobilans de biocarburants: une revue des controverses. Nat Sci Soc 16(4):337–347
Dorin B, Brunelle T, Crassous R, Gitz V, Hourcade J-C, Le Cotty T (2009) Compétition énergie-alimentation dans l'usage des sols. Eléments d'analyse et de modélisation des perspectives globales de développement des biocarburants. Cahiers de l'Énergie, Conseil français de l'Énergie, World Energy Council, Paris, pp 137–298
D4E (2006) Changement climatique et usages des terres: vers le modèle Nexus Land Use, Direction des études économiques et de l'évaluation environnementale, ministère de l'Écologie, du développement et de l'aménagement durables, collection Études et Synthèses, 122 p
EEA (2001) Scenarios as tools for international environmental assessments, European Environment Agency, Environmental Issue Report, 24, 31 p
EEA (2007) Land-use scenarios for Europe: qualitative and quantitative analysis on a European scale. European Environment Agency, Technical Report, 9, 75 p
Eshel G, Martin PA (2006) Diet, energy and global warming. Earth Interact 10(9):1–17
Falkenmark M (1995) Coping with water scarcity under rapid population growth. In: Conference of Ministers responsible for Water of the Southern Africa Development Community, 23–24 November 1995, Pretoria, South Africa
FAO (2000) The state of food and agriculture 2000, Lessons from the Past 50 Years, Agricultural and Development Economics Working Papers, 32, Food and Agriculture Organization, Rome, 313 p
FAO (2001) Climate variability and change: a challenge for sustainable agricultural production. In: Proceedings of the 16th Session of the Committee of Agriculture, 26–30 March 2001, Rome
FAO (2003a) In: Bruinsma J (eds) World agriculture: towards 2015/2030. A FAO Perspective. Earthscan Ltd., London, 444 p
FAO (2003b) FAO methodology for the measurement of food deprivation. Food and Agriculture Organization of the United Nations, Statistics division, Rome, 12 p
FAO (2004) Food security as Rural Development Strategy. In: Proceedings of the 28th FAO Regional Conference for Latin America and the Caribbean, 26–30 April 2004, Guatemala City, Guatemala

FAO (2008) Hunger on the rise, Briefing paper, 17 September, Food and Agriculture Organization of the United Nations, Rome, 6 p
FAO (2009) The state of food insecurity in the world, economic crisis. Impacts and lessons learned, Food and Agriculture Organization of the United Nations, Rome, 56 p
Fischer G, Velthuizen H van, Nachtergaele FO (2000) Global agro-ecological zones assessment: methodology and results, Interim Report 00-064. International Institute for Applied Systems Analysis, Laxenburg, 314 p
Fischer G, Shah M, Velthuizen H van, Nachtergaele FO (2001) Global agro-ecological assessment for agriculture in the 21st century. International Institute for Applied Systems Analysis, Laxenburg, 44 p
Fischer G, Velthuizen H van, Shah M, Nachtergaele FO (2002) Global agro-ecological assessment for agriculture in the 21st century: methodology and results, Research Report RR-02-02. International Institute for Applied Systems Analysis and Food and Agriculture Organization of the United Nations, Laxenburg, 119 p
Fischer G, Shah M, Tubiello FN, Velhuizen H van (2005) Socioeconomic and climate change impacts on agriculture: an integrated assessment 1990–2080. Philos Trans R Soc B 360(1463):2067–2083
Griffon M (2002) Révolution verte, Révolution doublement verte: quelles technologies, institutions et recherche pour les agricultures de l'avenir? Mondes en développement 1(117):39–44
Griffon M (2006) Nourrir la planète. Pour une révolution doublement verte. Odile Jacob, Paris, 455 p
Hoogwijk M, Faaij A, Eickhout B, de Vries B, Turkenburg W (2005) Potential of biomass for energy out to 2100, for Four IPCC SRES land-use scenarios. Biomass Bioenerg 29(4):225–257
Horrigan L, Lawrence RS, Walker P (2002) How Sustainable agriculture can address the environmental and human health harms of industrial agriculture. Environ Health Perspect 110(5):445–456
Hubert B, Brossier J, Caron P, Fabre P, de Haen H, Labbouz B, Petit M, Treyer S (2010) Forward thinking in agriculture and food. A platform for a dialogue to be continued. Perspective-Research, Cirad n. 6, Sept 2010, 4 p
IAASTD (2009) In: McIntyre BD, Herren HR, Wakhungu J, Watson RT (eds) Agriculture at a crossroads, global report. International Assessment of Agricultural Knowledge, Science and Technology for Development. Island Press, Washington, DC, 606 p
INSERM, TNS Healthcare, Roche (2003) ObEpi Roche, Enquête épidémiologique nationale sur l'obésité et le surpoids en France 2003, Une enquête INSERM/TNS HEALTHCARE SOFRES/Roche
INSERM, TNS Healthcare, Roche (2009) ObEpi Roche 2009, Enquête épidémiologique nationale sur le surpoids et l'obésité, Une enquête INSERM/TNS HEALTHCARE (KANTAR-HEALTH)/Roche, 52 p
IPCC (2000) In: Nakicenovic N, Swart R (eds) Special Report on Emissions Scenarios. A Special Report of Working Group III of the Intergovernmental Panel on climate change. Cambridge University Press, Cambridge, UK, 570 p
IPCC (2001) In: McCarthy JJ, Canziani OF, Leary NA, Dokken DJ, White KS (eds) Climate change 2001: impacts, adaptation and vulnerability. Contribution of Working Group II to the Third Assessment Report of the Intergovernmental Panel on climate change. Cambridge University Press, Cambridge, UK, 1 032 p
IPCC (2007a) In: Pachauri RK, Reisinger A (eds) Climate change 2007: Synthesis Report, Contribution of Working Groups I, II and III to the Fourth Assessment. Report of the Intergovernmental Panel on climate change, Geneva, 104 p
IPCC (2007b) In: Parry ML, Canziani OF, Palutikof JP, van der Linden PJ, Hanson CE (eds) Climate change 2007: impacts, adaptation and vulnerability. Contribution of Working Group II to the Fourth Assessment Report of the Intergovernmental Panel on climate change. Cambridge University Press, Cambridge, UK, 976 p
IPCC (2007c) In: Solomon S, Qin D, Manning M, Chen Z, Marquis M, Averyt KB, Tignor M, Miller HL (eds) Climate change 2007: the physical science basis. Contribution of Working

Group I to the Fourth Assessment Report of the Intergovernmental Panel on Climate Change. Cambridge University Press, Cambridge, UK, 996 p
Johnson FX, Chen Y, Zuzarte F (2008) Biofuels, land use, and sustainable development in Asia and Africa. UNEP/GRID-Arendal, Environ Poverty Times 5:16–17
Jones TW (2004) Using contemporary archaeology an applied anthropology to understand food loss in the American food system. Bureau of Applied Research in Anthropology, University of Arizona, 9 p
Just RE, Zilberman D, Hochman E (1983) Estimation of multicrop production functions. Amer J Agr Econ 65(4):770–780
Kader AA (2005) Increasing food availability by reducing post harvest losses of fresh produce. In: Mencarelli F, Tonutti P (eds) Proceedings of the 5th International Postharvest Symposium. Acta Horticulturae, 682, International Society for Horticultural Science
Kantor L, Lipton K, Manchester A, Oliveira V (1997) Estimating and adressing America's food losses. Food Rev 20(1):2–11
Kelly T, Yang W, Chen CS, Reynolds K, He J (2008) Global burden of obesity in 2005 and projections to 2030. Int J Obes 32(9):1431–1437
Koning NBJ, Ittersum MK van, Becx GA, Boekel MAJS van, Brandenburg WA, Broek JA van den, Goudriaan J, Hofwegen G van, Jongeneel RA, Schiere JB, Smies M (2008) Long-term global availability of food: continued abundance or new scarcity? NJAS Wagening J Life Sci 55(3):229–292
Ladha JK, Dawe D, Pathak H, Padre AT, Yadav RL, Bijay-Singh, Yadvinder-Singh, Singh P, Kundu AL, Sakal R, Ram N, Bhattarai EM, Gupta RK, Hobbs PR (2003) How extensive are yield declines in long-term rice-wheat experiments in Asia? Field Crops Res 81(2–3):159–180
Lahmar R (2006) Opportunités et limites de l'agriculture de conservation en Méditerranée. Les enseignements du projet KASSA. In: Arrue Ugarte JL, Cantero Martínez C (eds) Troisièmes rencontres méditerranéennes du semis direct. CIHEAM-IAMZ, Zaragoza, pp 11–18 (options méditerranéennes, série A., Séminaires Méditerranéens, 69)
Latham MC (1997) Human nutrition in the developing world. Food and Agriculture Organization of the United Nations, Rome, 515 p
Lobell DB, Burke MB, Tebaldi C, Mastrandrea MD, Falcon WP, Naylon RL (2008) Prioritizing climate change adaptation needs for food security in 2030. Science 319(5863):607–610
Luyten JC (1995) Sustainable world food production and environment, Internal Report 37. DLO-Inst. for Agrobiological and Soil Fertility Research, Wageningen, 127 p
MA (2005a) Ecosystems and human well-being: synthesis, millennium ecosystem assessment. Island Press, Washington, DC, 155 p
MA (2005b) In: Carpenter SR, Pingali PL, Bennett EM, Zurek MB (eds) Ecosystems and human well-being: scenarios, volume 2, Millennium ecosystem assessment. Island Press, Washington, DC, 560 p
MA (2005c) In: Hassan R, Scholes R, Ash N (eds) Ecosystems and human well-being: current state and Trends, volume 1, Millennium ecosystem assessment. Island Press, Washington, DC, 917 p
Malassis L, Padilla M (1986) Économie agroalimentaire. L'économie mondiale (III). Cujas, Paris, 449 p
McCalla AF, Revoredo CL (2001) Prospects for global food security. A critical appraisal of past projections and predictions. International Food Policy Research Institute. 2020 Discussion Paper, 35, Washington, DC, USA, 71 p
McNeely J, Scherr S (2003) Ecoagriculture: strategies to feed the world and to save biodiversity. Island Press, Washington, DC, 290 p
Meadows DH, Meadows DL, Randers J, Behrens WW III (1972) The limits to growth. Universe Books, New York, 205 p
Ministère de la santé et des solidarités (2006) Deuxième Programme national nutrition santé 2006–2010. Actions et mesures, 51 p
Mishra S (2007) A Brief History of Production Functions. Munich Personal RePEc Archive Paper 5254, North-Eastern Hill University, Shillong (India), 9 October, 23 p

Murdiyarso D (2000) Adaptation to climatic variability and change: Asian perspectives on agriculture and food security. Environ Monit Assess 61(1):123–131
Myers N (1991) Population, resources and the environment: the critical challenge. United Nations Fund for Population Activities, New York, 154 p
OECD (2008) Promoting sustainable consumption. Good practices in OECD countries. Organisation for Economic Co-operation and Development, 62 p
Olshansky SJ, Passaro DJ, Hershow RC, Layden J, Carnes BA, Brody J, Hayflick L, Butler RN, Allison DB, Ludwig DS (2005) A potential decline in life expectancy in the United States in the twenty-first century. N Engl J Med 352(11):38–45
Parry ML, Rosenzweig C, Iglesias A, Livermore M, Fischer G (2004) Effects of climate change on global food production under SRES. Emissions and socio-economic scenarios. Global Environ Chang 14:53–67
Pimentel D, Pimentel M (1996) Food, Energy and Society, revised edition. CO University Press of Colorado, Niwot, 363 p
Popkin BM (1993) Nutritional patterns and transitions. Popul Dev Rev 19(1):138–157
Popkin BM (2008) Will China's nutrition transition overwhelm its health care system and slow economic growth? Health Aff 27(4):1064–1076
Rastoin J-L (2007) Risks and food safety in a context of globalization: towards a political and strategic approach. In: Hervieu B (eds) Mediterra, identity and quality of mediterranean foodstuffs. les Presses de Sciences Po, Paris, pp 27–68
Rosegrant MW, Paisner MS, Meijer S, Witcover J (2001) Global food projections to 2020. Emerging trends and alternative futures. International Food Policy Research Institute, Library of Congress Cataloging in Publication Data, 206 p
Rosegrant MW, Cai X, Cline SA (2002) World water and food to 2025: dealind with scarcity. International Food Policy Research Institute, Library of Congress Cataloging in Publication Data, 322 p
Ruhm CJ (2007) Current and future prevalence of obesity and severe obesity in the United States. Forum Health Econom Policy 10(2):1–26
Salanave B, Peneau S, Rolland-Cachera MF, Hercberg S, Castetbon K (2008) Stabilization of overweight prevalence in French children between 2000 and 2007. Int J Paediatr Obes 4(2):66–72
Sebillote M (Dir.) (2001) Les protéines végétales et animales: enjeux de société et défis pour l'agriculture et la recherche. Institut national de la recherche agronomique, Paris, 195 p
Smeets EMW, Faaij APC, Lewandowski IM, Turkenburg WC (2007) A bottom-up assessment and review of global bioenergy potentials to 2050. Prog Energy Combust Sci 33:56–106
Smil V (2000) Feeding the world: a challenge for the twenty-first century. MIT Press, Cambridge, 362 p
Steeg JA van de, Herrero M, Kinyangi J, Thornton PK, Rao KPC, Stern R, Cooper P (2009) The influence of climate variability and climate change on the agricultural sector in East And Central Africa. Sensitizing the ASARECA Strategic plan to climate change. Research Report 22, ILRI (International Livestock Research Institute), Nairobi, Kenya. ICRISAT (International Crop Research Institute for the Semi-Arid Tropics), Nairobi, Kenya, and ASARECA (Association for Strengthening Agricultural Research in Eastern and Central Africa), Entebbe, Uganda
Steinfeld H, Gerber P, Wassenaar T, Castel V, Rosales M, Haan C (2006) Livestock's long shadow, environmental issues and options. Food and Agricultural Organization of the United Nations, Rome, 390 p
Swinburn B, Egger G (2002) Preventive strategies against weight gain and obesity. Obes Rev 3(4):289–301
Taithe A (2007) L'eau, facteur d'instabilité en Chine. Perspectives pour 2015 et 2030. Fondation pour la Recherche Stratégique, Recherches et documents, 53 p
Thornton PK, Jones PG, Owiyo T, Kruska RL, Herrero M, Kristjanson P, Notenbaert A, Bekele N, Omolo A, with contributions from Orindi V, Otiende B, Ochieng A, UNDP (2005) Human Development Report 2005, international cooperation at a crossroads: aid, trade and security in an unequal world. United Nations Development Programme, 372 p

UNDP (2005) Human Development Report 2005, international cooperation at a crossroads: aid, trade and security in an unequal world. United Nations Development Programme, 372 p
UNEP (2007) Global Environment Outlook 4 (GEO-4). Environment for Development, United Nations Environment Programme, 540 p
USEN (Unité de surveillance et d'épidémiologie nutritionnelle) (2007) Étude nationale nutrition santé (ENNS, 2006). Situation nutritionnelle en France en 2006 selon les indicateurs d'objectif et les repères du Programme national nutrition santé (PNNS). Institut de veille sanitaire, Université de Paris xiii, Conservatoire national des arts et métiers, 74 p
Watson RT, Zakri AH (2005) Foreword. In: Carpenter SR, Pingali PL, Bennett EM, Zurek MB (eds) Ecosystems and human well-being: scenarios, Volume 2. Island Press, Washington, DC, p xix (Millennium Ecosystem Assessment)
WHO and FAO (2003) Diet, nutrition and the prevention of chronic diseases. WHO Technical Report Series, Report of a Joint World Health Organisation/Food and Agriculture Organisation Expert Consultation, 28 January–1 February 2002, Geneva, 149 p
Wirsenius S (2003) Efficiencies and biomass appropriation of food commodities on global and regional levels. Agric Syst 77(3):219–255
World B (2008) Agriculture for development, World Development Report 2008. The World Bank, Washington, DC, 365 p
WRAP (2008) The food we waste. Food Waste Report, Waste and Resource Action Programme, 236 p
Zimmer D, Renault D (2003) Virtual water in food production and global trade: review of methodological issues and preliminary results. In: Hoekstra AY (eds) Proceedings of the International Expert Meeting on Virtual Water Trade, 12–13 December 2002, Delft, The Netherlands, Value of Water Research Report Series 12, UNESCO-IHE, pp 93–109

References to web sites or to data bases

FAO (2003) Food balance sheet. A handbook, Food and Agricultural Organization of the United Nations. http://www.fao.org/docrep/003/X9892E/X9892E00.HTM
FAO (2006) Faostat, internet web portal and database as on 30 June 2006, Food and Agricultural Organization of the United Nations. http://faostat.fao.org/site/395/default.aspx. Accessed 30 June 2006
FAO, IIASA (2000) Global agro-ecological zones, Food and Agriculture Organization of the United Nations, International Institute for Applied Systems Analysis, http://www.fao.org/AG/agL/agll/gaez/index.htm. Accessed 22 June 2007
FAOSTAT (2008) Fisheries and aquaculture information and statistics service, Food and Agriculture Organization of the United Nations. http://www.fao.org/fishery/statistics/global-aquaculture-production/query/en
Kaas R (2006) The seaweed resources of france. In: Critchley AT, Ohno M, Largo DB (eds) World seaweed resources. An Authoritative Reference System 2006. A Multimedia, Interactive DVD ROM, ETI Bioinformatics, Amsterdam
UN (2006) World population prospects: the 2006 revision population database
UN (2007) World urbanization prospects: the 2007 revision population database
USDA (2006) USDA national nutrient database for standard reference. Release 19, U.S. Department of Agriculture. http://www.nal.usda.gov/fnic/foodcomp/search/. Accessed 20 Aug 2007

MIX
Papier aus verantwortungsvollen Quellen
Paper from responsible sources
FSC® C105338

If you have any concerns about our products,
you can contact us on
ProductSafety@springernature.com

In case Publisher is established outside the EU,
the EU authorized representative is:
Springer Nature Customer Service Center GmbH
Europaplatz 3, 69115 Heidelberg, Germany

Printed by Libri Plureos GmbH
in Hamburg, Germany